Differential Modules over Differential Rings

Differential Modules over Differential Rings provides an introduction and reference for researchers in commutative and differential algebra and could be used as the basis for a graduate course or seminar. The book is best suited to an audience for whom the terminology of rings, modules, homomorphisms, and categories is already familiar. Although the topic is rooted in differential algebra, and the book should be of interest to workers in that area, no particular prior knowledge of differential algebra is assumed. When it is necessary to use specialized results from differential algebra, especially Picard—Vessiot theory, the necessary definitions and theorems are supplied.

Features

- Collects the basic definitions and results about differential modules in one convenient reference with uniform notation.
- Accessible to readers who don't have extensive specialized knowledge of differential algebra or commutative ring theory.
- The first book of its kind dedicated exclusively to the topic in this generality.
- Presents new formulations of previously published work as well as new results not previously published.

Andy R. Magid is George Lynn Cross Professor of Mathematics Emeritus at the University of Oklahoma whose faculty he joined in 1972. He holds the B.A. and PhD degrees in Mathematics from the University of California and Northwestern University, respectively. He was in the inaugural class of Fellows of the American Mathematical Society.

Monographs and Research Notes in Mathematics

Series Editors: John A. Burns, Thomas J. Tucker, Miklos Bona, and Michael Ruzhansky

This series is designed to capture new developments and summarize what is known over the entire field of mathematics, both pure and applied. It will include a broad range of monographs and research notes on current and developing topics that will appeal to academics, graduate students, and practitioners. Interdisciplinary books appealing not only to the mathematical community, but also to engineers, physicists, and computer scientists are encouraged.

This series will maintain the highest editorial standards, publishing well-developed monographs as well as research notes on new topics that are final, but not yet refined into a formal monograph. The notes are meant to be a rapid means of publication for current material where the style of exposition reflects a developing topic.

Completely Regular Codes in Distance Regular Graphs
Patrick Solé and Minjia Shi

Non-Newtonian Sequence Spaces with Applications
Feyzi Başar and Bipan Hazarika

Spectral Theory for Linear Operators
Bilel Krichen

Convex Analysis in Polynomial Spaces with Applications
Domingo García, Mingu Jung, Manuel Maestre, Gustavo A. Muñoz Fernández and Juan B. Seoane Sepúlveda

Modelling Order and Disorder
Integro–Differential Nonlinear Equations
Miroslaw Lachowicz

Free Random Variables
Free Distributions Dictated by the Semicircular Law
Ilwoo Cho

Inequalities and Integral Operators in Function Spaces
Erlan Nursultanov

Monomial Algebras
Rafael H. Villarreal

Finite Element Methods for Eigenvalue Problems, Second Edition
Jiguang Sun and Aihui Zhou

Differential Modules over Differential Rings
Andy R. Magid

For more information about this series please visit: www.routledge.com/Chapman--Hall-CRC-Monographs-and-Research-Notes-in-Mathematics/book-series/CRCMONRESNOT

Differential Modules over Differential Rings

Andy R. Magid

CRC Press
Taylor & Francis Group
Boca Raton London New York

CRC Press is an imprint of the
Taylor & Francis Group, an **informa** business
A CHAPMAN & HALL BOOK

Designed cover image: Andy R. Magid

First edition published 2026
by CRC Press
2385 NW Executive Center Drive, Suite 320, Boca Raton FL 33431

and by CRC Press
4 Park Square, Milton Park, Abingdon, Oxon, OX14 4RN

CRC Press is an imprint of Taylor & Francis Group, LLC

ISBN: 978-1-032-58810-0 (hbk)
ISBN: 978-1-032-58928-2 (pbk)
ISBN: 978-1-003-45216-4 (ebk)

DOI: 10.1201/9781003452164

Typeset in Latin Modern font
by KnowledgeWorks Global Ltd.

Publisher's note: This book has been prepared from camera-ready copy provided by the author.

Contents

Preface

"HISTORY OF THE PROBLEM"

Morris Orzech, of Queens University, where master's students typically write and orally defend a thesis, was meeting with one such student to discuss the student's presentation part of the defense. The student offered that he would like to present the history of the problem. Orzech agreed. At the exam, the student began with something like "when Professor Orzech first gave me this problem, I wasn't sure what to do, then I tried [so and so] ..." This is not the customary usage of the phrase "history of the problem" among mathematicians. Over the years,[1] however, when I have had occasion to want to say something personal about a mathematical development, I have found it convenient to refer to "the history of the problem in the sense of Orzech's student". This is one of those occasions.

About two decades ago, Lourdes Juan and I began talking about investigating K-theory for the category of finite-dimensional differential modules over a differential field F (always of characteristic zero and with an algebraically closed field of constants C, so Picard–Vessiot theory could be used). For reasons I no longer remember, and which are probably not true, we decided that this would not be an interesting

[1] I heard Morris tell this story half a century ago.

project: the K groups in question would just be free groups on the isomorphism classes of the simple modules, and saying anything in general about what the simple modules are over an arbitrary F seemed hopeless (and probably is). On the other hand, we could also see that if instead of looking at differential F modules, we looked instead at central simple F algebras, we could see some interesting things to investigate. Since, as Hyman Bass had shown long ago, the Brauer group is the reduced K_0 of the category of Azumaya algebras (reduced by modding out the classes of endomorphism rings of projective modules), we decided to look in particular at the differential analogue, which we called the differential Brauer group. (I should point out here that our 2008 account of this investigation, published in the *Proceedings* of the American Mathematical Society, has some errors recently pointed out and corrected, as far as possible, by Wang and Srinivasan.) Lourdes had an East Coast sabbatical about this time and took the opportunity to present our results in various places, including New York, where my old friend and coauthor – and Brauer group expert – Ray Hoobler learned about them. Over the years I, Lourdes, Ray, and various subsets, discussed extending the work Lourdes and I had done to differential Brauer groups in the partial derivative case, among others. The sticking point in the partial derivative case was the following: for the differential Brauer group of an ordinary differential field F, the fact that any derivation of the center of a central simple algebra extends to a derivation of the algebra means that the differential Brauer group maps subjectively to the ordinary Brauer group. Although we could define the differential Brauer group easily enough in the partial derivative case, for surjectiveity onto the ordinary Brauer group, we would need to know that we could extend commuting derivations of the center of a central simple algebra to *commuting* derivations of the algebra.

There are several proofs of the extension of a (single) derivation of the center of a central simple algebra to a derivation of the algebra. What I think is the earliest, and certainly in my view the most explicit, is due to Gerhard Hochschild. Having n commuting derivations on F which are constant on a given algebraically closed characteristic zero subfield C is like having the abelian Lie algebra $\mathcal{L}$ of rank n over C act on F as derivations. Extending the derivations to commuting derivations of a central simple F algebra A is equivalent to extending the action of $\mathcal{L}$ to A acting as derivations. One could ask the same question when F is replaced by a commutative ring R (still an algebra over the field C) and an Azumaya R algebra A. Actually, I did ask that in depth of Hochschild himself during a sabbatical at Berkeley in the spring quarter of 1980. My interest was in the case where $\mathcal{L}$ was the Lie algebra of an algebraic group G over C, where G acted rationally on R. I wanted to extend the G action to an Azumaya R algebra A. I thought there might be a chance that if the $\mathcal{L}$ action could be extended to A then so could the G action. Although my conversations with Hochschild on this matter were certainly stimulating, we never resolved anything. (I asked the question, strictly in the group form, at math receptions for years afterward. My recollection is that Canepeel provided a counterexample.)

I was interested in the algebraic group question because I had taken an interest in commutative rings with algebraic group action. At the time (late 1970s) the celebrated Hochster-Roberts theorem had recently been proved: if R is a regular commutative ring on which the linearly reductive algebraic group G acts rationally, then the ring of invariants R^G is

Cohen–Macauley. I had the idea (spoiler alert: I couldn't make it work) that since both "regular" and "Cohen–Macauley" are properties of the category of modules; more precisely, homology of injective resolutions in the category of modules, I could use the category of R modules which were also rational G modules, called $R \cdot G$ modules, to mediate between the categories of R^G modules and the category of R modules, since there are functors from R^G modules to $R \cdot G$ modules ($M \mapsto R \otimes_{R^G} M$) and from $R \cdot G$ modules to R modules (forget the G action) and R^G modules ($N \mapsto N^G$) and if G is linearly reductive this latter is exact. In the fall semester of 1989, on the same sabbatical that took me to Berkeley for the spring quarter, I was invited to the University of Virginia, where I had access to Brian Parshall and Leonard Scott, certainly as expert in $R \cdot G$ modules over any field as could be found, as well as commutative algebraist Doug Costa. I worked out plenty of machinery, which I eventually published, especially injective hulls, Matlis theory of injectives, and Bass numbers, for $R \cdot G$ modules, although it never yielded the Hochster-Roberts theorem. My student Ida Doraiswamy, whose PhD was granted in 1980, although her work was completed earlier, had shown in her thesis that when R is G simple, in the sense of having no G stable ideals, then finitely generated $R \cdot G$ modules were R projective, I had asked Ida to look at $R \cdot G$ modules (which we called *modules over a *ring) after having read the work of Fossum and Foxby about $\mathbb{Z}$ graded modules over a graded commutative ring (which they called *modules) and realizing that $\mathbb{Z}$ grading for a $\mathbb{C}$ algebra R was the same as a rational G_m action on R. (Hans Foxby was visiting in Norman, which he did regularly for family reasons, and called my attention to that work but it was the end of that visit and we didn't have time to discuss it).

Much later (about 15 years ago) I replicated much of the $R \cdot G$ module theory for differential modules over a differential ring. Hewing close to the parameters of the earlier work with algebraic group rational action on the ring and modules, I restricted attention to D finite derivations on the ring and modules. I did find that when the ring was differentially simple (had no proper differential ideals) then finitely generated differential modules were projective. (Meanwhile Yves André had established that without needing D finite.) Perhaps because of this, when Lourdes and I resumed our investigations of differential modules about a decade ago, we focused on those finitely generated and projective as modules over the ring, and soon discovered that every such projective module over the ring carried differential structures. We also found a criteria to decide which differential finitely generated projective R modules were of the form $R \otimes_{R^D} P$ for some finitely generated projective R^D module P. We were still stymied about classification: using differential K_0 still introduced a relation we couldn't control

Meanwhile, 2019 I had to spend several spring months in New York on family business so Ray Hoobler (by then retired from CUNY but still a regular presence in the Graduate Center Mathematics Department) and I, hoping to do some research which would lead to a paper suitable for publication in the *Proceedings* of the American Mathematical Society roughly fifty years after we had a joint paper there, decided to look at the category of differential Azumaya algebras over a differential commutative ring and its K_0. We didn't finish, but we had some ideas, and I had several more occasions to come to New York, which by the time of our final conversations in mid–February 2020 had centered on monoids, rather than groups, as the classifying algebraic structure. Ray, unfortunately, was a victim shortly after of the COVID-19 epidemic then ravaging New York, but I was able to work through our research plan, which was eventually published in the *Proceedings* under the title "Differential Brauer Monoids" in 2023, only one year later than the target semi-centennial. The same classification technology applies to categories of differential finitely generated projective modules over a differential commutative ring, which Lourdes and I did, although the classifying monoid doesn't have as nice a name (we called it the "differential projective class monoid").

I also want to say something about my entry into differential algebra. In the spring of 1992, I was on sabbatical at the Hebrew University. I think I was asked if I wanted to offer a course. I had always been intrigued by differential Galois theory, so I said I would lecture on that, with the idea that I would more or less follow Kaplansky's little book. Very quickly I found myself working out the differential algebra of Picard–Vessiot rings which (among other things) are commutative rings on which the differential Galois group of the Picard–Vessiot extension acts rationally without any stable ideals. (I defined the rings in terms of complete sets of solutions of a differential equation.) Set up properly this eventually yields the fundamental theorem of differential Galois theory. When I returned to Oklahoma I reprised my lectures as a seminar for my department, and then wrote them up. The American Mathematical Society published these in their University Lecture Series in 1994. Soon after (1996), Daniel Bertrand reviewed them for the *Bulletin* of the American Mathematical Society. In addition to some kind words about my book, Bertrand's review gave a concise exposition of the differential module based (as opposed to solutions of differential equations based) approach to differential Galois theory that uses Grothendieck's theory of Tannakian categories. Now I was no stranger to Tannakian categories: the work of Hochschild and Mostow that I had investigated that had led to my 1980 visit to Hochschild explicitly referred to Tannaka Duality, and I knew even then that what I was looking at was a Tannaka Duality group in the sense of Grothendieck. However, that it was a tool for differential Galois theory was news to me (even though Deligne had pointed it out in 1990). Picard–Vessiot rings play no explicit role in the Tannakian approach. Nonetheless, if one has already worked out differential Galois theory with Picard–Vessiot rings, one can show that the category of differential modules is equivalent to the category of modules for the differential Galois group, which is sort of the Tannakian approach. I took the opportunity of a lecture at a Fields Institute 2002 conference on category theory to work out the details of that equivalence using Picard–Vessiot rings.

There is another plot twist to add to this narrative, for its own sake and for use below. I was at the 2002 Fields conference at the invitation of one of the organizers, George Janelidze. Janelidze is really the creator of categorical Galois theory. My 1969 PhD thesis is about the separable Galois theory of commutative rings, and I had given an account of that subject in a 1974 book of that title. There is a Galois correspondence theorem for appropriate extensions of commutative rings using a certain profinite groupoid: certain closed subgroupoids correspond to intermediate extensions. In the book, repeating an argument that I had previously published in the *Transactions* of the American Mathematical Society, I claimed to prove that closed subgroupoids of profinite groupoids were inverse limits of finite groupoids. This is false. The correspondence theorem is actually between intermediate extensions and subgroupoids which are inverse limits of finite groupoids. I had been circulating this correction (and some typo corrections) informally for many years when George invited me to join a paper with him and Aurelio Carboni which provided an opportunity to put this in the archival referred literature. And George also thought, correctly, that I would benefit from participating in the Fields conference.

I hope the serendipitous nature of my investigations of differential modules has come through in the preceding remarks. When we write mathematics, we present the work orderly and logically. No discussion of the dead ends, false conjectures, and unsuccessful proofs ever appears in the finished paper. When we write research proposals, we similarly present an orderly and logical agenda (or at least I did. I was also advised to never propose research you didn't already know how to do, which I didn't do.) Again speaking for myself, the actual research work is never that neat. Similarly, when I read the narrative just provided here, there seems to be more pattern in retrospect than it seemed as it came to pass.

THIS BOOK

Differential algebra is a vast topic. Even commutative differential algebra is a vast topic. For example, everything about differential fields falls under this heading. Modules over a commutative differential ring is also a vast topic. Among the things that could be considered under that heading (and are not looked at herein) are differential dimension theory or decomposition theory of differential ideals (the analog of primary decomposition). What this book is about is commutative rings with a derivation, and the category of modules with derivations over those rings, guided by three unstated general goals. The first concerns looking for which differential modules are given by constants, or what has to be done to convert a given differential module to one of this form. The second looks at what homological algebra, generally in the form of injective resolutions, means for differential modules. The third is trying to classify differential modules up to isomorphism. In all cases, its not just differential modules but the homomorphisms between them (although not always differential homomorphisms) that are in the forefront. To be concrete:

We begin, with Chapter 1, with some basics about a commutative differential ring, including the interpretation of derivations as homomorphisms to dual numbers. Basics used in subsequent chapters are covered, including differential ideals and derivations of polynomial rings. Differential modules are introduced and so are the differential structures on algebraic constructions with differential modules: homomorphism modules, duals, tensor products, and symmetric algebras. Also considered are the differential modules whose underlying module is a finitely generated free module and how such modules are given by matrices. Chapter 2 is about differential modules over differential fields for which the Picard–Vessiot differential Galois is available: characteristic zero differential fields F with an algebraically closed field of constants C. After reviewing the Galois theory, including the Picard–Vessiot closure F^{PV} and its Picard–Vessiot ring R^{PV}, as well as the proalgebraic differential Galois group Π of F^{PV} over F. It is shown how R^{PV} can be used to set up a duality between finite-dimensional differential F modules and finite dimensional (over C) rational Π modules. The chapter explains how trivializing the differential F modules (that is, solving the differential equations they represent) comes about. Chapter 3 considers differential commutative rings R which are algebras over a differential field F (usually a field of the type considered in Chapter 2), and differential R modules M. It focuses on the D finite case: when every element of R. and every element of M satisfies a monic linear homogeneous differential equation with coefficients in F, A good example of such an R is R^{PV} which is studied carefully. Injective resolutions and basic homology of these rings are investigated, especially H^1 and extensions. Chapter 4 takes up differential modules over a differential ring R which are finitely generated and projective as modules over the ring. As is shown, every finitely generated projective R module has a differential structure. This important observation is actually shown twice, first conceptually using dual numbers and then more explicitly, using projective coordinate systems and idempotent matrices. Another important point covered in this chapter is how differential finitely generated projective R modules of the form $R \otimes_{R^D} P$, with P finitely generated and projective over R^D can be detected. Chapter 5, the final one, looks at the classification of differential finitely generated projective modules over a differential commutative ring R. Of course, they are classified by isomorphism classes, and the isomorphism classes form a monoid under the operation induced from $\oplus$, but the goal is something more manageable. Following some of the inspirations of K-theory, the classification studied is the monoid which, basically, is the most general monoid (1) to which the monoid of isomorphism classes maps homomorphically; and (2) in which the isomorphism classes of all the modules of the form $R \otimes_{R^D} (R^D)^n$ go to the identity.

ACKNOWLEDGEMENTS

I noted above how our published mathematics tends to obscure the messier aspects of our discoveries. Hopefully, all the papers with which I was involved on the topics covered in this book presented the mathematics cleanly enough. However, no matter how well organized the individual papers were, taken as a whole, there were always things I learned later that could have been said clearer, or more elegantly, and with better worked-out definitions and statements. The opportunity to "get it right", or at least get it in better shape for the record and for subsequent mathematicians who might want to further investigate the topic, is one that authors always benefit from, and hopefully also the larger mathematical community too. Here's how that came about in this case. I mentioned above that *The Separable Galois Theory of Commutative Rings*, published by Marcel Dekker in 1974, had an error I needed to correct. I can't remember when I contracted to provide a revised edition, but I was certainly tardy in submitting the manuscript, which was published by CRC Press in 2014 as the second (and incidentally fortieth anniversary) edition, I didn't think that I had any more to say on the subject (although others have), so when Taylor and Francis approached me about a third edition, I of course declined. However, that did make me think that perhaps a book about my work with differential modules might be possible, and drafting a proposal was encouraging. I am grateful to Taylor and Francis, in particular Callum Fraser, for reaching out which started this project in motion. Just as I was reaching the stage of drafting the chapter on differential modules over fields, invitations from Ivan Tomasic (Queen Mary University of London) and Anand Pillay

(University of Notre Dame) provided occasions to think deeply about the topic, expert collaborators to talk it over, and a venue to present in. I have already mentioned the collaborations with Lourdes Juan and the late Ray Hoobler from which much of the original work covered in this book stems. It remains to thank my family: the Dekker volume is dedicated to my wife Carol; the preface to the CRC second edition adds my children Sarah and Sam; and for this book, I am proud to include my grandchildren Levi and Charlee.

Andy Magid

Norman Oklahoma 2025

CHAPTER 1

Differential Rings

1.1 DEFINITIONS

Let R be a commutative ring. A *derivation* D of R is an additive homomorphism $D : R \to R$ satisfying the Leibnitz rule with respect to multiplication:

$$\text{For } a, b \in R,\ D(ab) = D(a)b + aD(b).$$

A pair (R, D) where R is a commutative ring and D is a derivation of R is called a *differential ring*. Typically we refer to such a pair by its first element and say "R is a differential ring" and then denote the derivation by D_R, or just D. Any given ring R can have many derivations and so be a differential ring in many ways. We note that nothing in the definition uses that R is commutative or even associative. If we need to refer to a non-commutative ring with a derivation, we will always specify that it is a non-commutative differential ring; differential rings without further qualification are always commutative.

Let R be a differential ring. Apply the Leibnitz rule when $a = b = 1$:

$$1 = 1 \cdot 1 \text{ implies } D(1) = D(1 \cdot 1) = D(1) \cdot 1 + 1 \cdot D(1) = D(1) + D(1),$$

DOI: 10.1201/9781003452164-1

so $D(1)=0$.

More generally, if $e \in R$ is idempotent,

$$e=e\cdot e \text{ implies } D(e)=D(e\cdot e)=D(e)\cdot e+e\cdot D(e), \text{ so } D(e)=2eD(e);$$

$$\text{multiplying by } e \text{ gives } eD(e)=2eD(e), \text{ so } eD(e)=0,$$

$$\text{and hence } D(e)=2eD(e)=0.$$

The situation with nilpotent elements is more complicated. We will see below an example of a differential ring and a nilpotent element whose derivative is not nilpotent. On the other hand, we will also see that under quite general circumstances (when R is a $\mathbb{Q}$ algebra suffices), nilpotents have nilpotent derivatives.

If $a \in R$ is invertible with inverse b, so $1=ab$, $0=D(1)=D(ab)=D(a)b+aD(b)$, then $aD(b)=-D(a)b$ and multiplying by b gives $D(b)=-D(a)b^2$.

For a differential ring R and an element $f \in R$, and a natural number n, an induction argument shows that $D(f^n)=nf^{n-1}D(f)$.

An element of R that lies in the kernel of D is called a *constant*. The set of constants of R is denoted R^D. It is a subring of R.

Every commutative ring has at least one derivation, namely the zero map; we will denote the corresponding differential ring as $(R,0)$. Sometimes this is the only derivation. If $R=\mathbb{Z}$ and D is a derivation of R then

$$\text{Since } D(1)=0, \text{ for } n\in\mathbb{Z}, D(n)=nD(1)=0.$$

Thus the unique derivation of $\mathbb{Z}$ is 0. If R is a finite field of characteristic p with $q=p^m$ elements, then since $x^q=x$ for $x \in R$, then for any derivation D of R, $D(x)=D(x^q)=qx^{q-1}D(x)$. Because the characteristic p is a factor q, $qx^{q-1}D(x) = 0$, so $D(x) = 0$. Thus the unique derivation of a finite field is the zero derivation. So any ring with two or three elements has only the zero derivation. But as we show below, there is a ring with four elements that has a non-zero derivation.

In general, of course, rings have many derivations. As we see shortly, a countable ring can have uncountably many. To construct such examples and to understand derivations in general, we will use dual numbers.

But first, a comment about origins. The Liebnitz rule, or product formula, of single-variable differential calculus, comes from the following reasoning: Suppose $f(x)$ and $g(x)$ are functions, and at the point x_0 let at and bt be the best linear approximations to f and g at x_0. Symbolically, $f(x) \sim f(x_0) + a(x - x_0)$ and $g(x) \sim g(x_0) + b(x - x_0)$, so $f(x)g(x) \sim (f(x_0) + a(x - x_0)(g(x_0) + b(x - x_0)) = f(x_0)(g(x_0) + (ag(x_0) + f(x_0)b)(x - x_0) + ab(x - x_0)^2$, which suggests that $(af(x_0) + g(x_0)b)t$ is the best linear approximation to $f(x)g(x)$ at x_0. It is, of course, easy to make this reasoning rigorous, part of which involves moving "has a best linear approximation at x_0" to the hypotheses and including "$f(x)g(x)$ has a best linear approximation at x_0" as a conclusion.

Rings of (infinitely) differentiable functions on C^∞ manifolds, or rings of complex analytic functions on complex analytic manifolds, are thus basic examples of differential rings. It is not necessary to consider all such functions on a given manifold: while the ring of all entire functions on the complex plane is an interesting differential ring (see Section 4.7), so is its subring consisting of the polynomials in the identity function. Studying differential rings in the abstract can be viewed as isolating the algebraic properties of rings of differentiable functions, and this certainly seems to describe the origin of the subject.

There is another source of examples, though, that should also be kept in mind, even if historically they were not considered as much. By a *dynamical system* we mean a set X (typically an analytic manifold or an algebraic variety), which is regarded as the set of *states* of the system, together with an action of the scalars, regarded as time, on X such that for $x \in X$ and a scalar t, $t \cdot x$ is regarded as the state the system is in if it were in state x and t units of time passed. It is required (i) that $0 \cdot x = x$ for all x (nothing happens if no time passes) and (ii) that $s \cdot (t \cdot x) = (s+t) \cdot x$ (if it starts in state x, then t units of time pass, and then s additional units of time pass, this is the same state it would be in starting at x and letting $s+t$ units of time pass). If f is a function on X and t is a scalar, we define the function $f \cdot t$ by the formula $f \cdot t(y) = f(t \cdot y)$.

If the variable really is time, then the scalars are the field of real numbers. For exposition, and with a view to abstraction, we will denote the scalars by the field F and their additive group by $(F, +)$. The two conditions (i) and (ii) then assert that $x \mapsto t \cdot x$ is a group action of $(F, +)$ on X and that $f \mapsto f \cdot t$ is a group action of $(F, +)$ on the functions on X. Now let R be a ring of functions on X such that if $f \in R$ and $t \in F$ then $f \cdot t \in R$. Let us assume further that R is a $\mathbb{C}$ algebra and that for $f \in R$ the $\mathbb{C}$-span of $\{f \cdot t | t \in F\}$ is finite dimensional. Note that this span is stable under the action of $(F, +)$ and hence affords a finite-dimensional linear representation of $(F, +)$. We further restrict to the case where F is either $\mathbb{R}$ or $\mathbb{C}$ so that $(F, +)$ is either a one-dimensional real Lie group or a one-dimensional complex analytic group. In either case we further assume that for every $f \in R$ the associated finite-dimensional representation is either C^∞ or analytic. As we will review below, when a

Lie group acts linearly on a vector space, its Lie algebra also acts on the vector space. These actions combine to give an action of the Lie algebra of $(F,+)$ on R – as derivations. This can be thought of as the instantaneous transformation of the dynamical system under an infinitesimal time unit. Since $(F,+)$ is one dimensional, so is its Lie algebra, and if D is a basis of that Lie algebra, (R,D) is then a differential ring. In many cases, for example, if X is an affine open subset of a complex affine space and the action is algebraic so that R can be taken as the coordinate ring of X, then D determines the action. Many physical systems can be so modeled. Understanding the differential ring amounts to understanding the dynamical system.

In summary, a differential ring is an abstraction of an instantaneous dynamical system.

1.2 EXTENSIONS

If R and S are both differential rings, and $g : R \to S$ is a ring homomorphism such that $D_S(g(x)) = g(D_R(x))$ for all $x \in R$, then g is said to be a *differential ring homomorphism.* If g is the inclusion, or can be regarded as the inclusion, then R is said to be a *differential subring* of S, and S is said to be a *differential ring extension* of R.

Sometimes it is convenient to speak about a derivation from a ring into another ring. That is, if we have commutative rings S and T and a commutative ring homomorphism $h : S \to T$ then an additive homomorphism $\delta : S \to T$ that satisfies a Liebnitz rule: $\delta(ab) = \delta(a)h(b) + h(a)\delta(b)$ is called a *derivation from* S *to* T; if we need to specify h, we call δ an h-derivation from S to T.

For differential rings in general, we consider the dual numbers construction, which allows derivations to be viewed as certain kinds of ring homomorphisms. Since various ring extensions, in particular polynomial rings and rings of fractions, are characterized by their homomorphism properties, this construction will allow us to carry those properties over to derivations.

The ring of *dual numbers* over R is the ring $R[X]/(X^2)$. Let ϵ stand for the class of X in this quotient. Then $R[X]/(X^2) = R + R\epsilon = R[\epsilon]$; the sum is direct and $R^2 \to R + R\epsilon$ by $(a,b) \mapsto a + b\epsilon$ is a bijection. Using this bijection, the ring homomorphism $R[X]/(X^2) \to R[X]/(X)$ becomes $\pi : R[\epsilon] \to R$ by $a + b\epsilon \mapsto a$. Moreover, $R[\epsilon]$ is an R algebra under the map $\iota : R \to R[\epsilon]$ by $a \mapsto a + 0\epsilon$. If $g : R \to S$ is a ring homomorphism, we let $g[\epsilon]$ be the ring homomorphism $R[\epsilon] \to S[\epsilon]$ by $a + b\epsilon \mapsto g(a) + g(b)\epsilon$. This is compatible with the maps ι and π: the diagrams commute.

$$\begin{array}{ccc} R & \xrightarrow{\iota} & R[\epsilon] \\ {\scriptstyle g}\downarrow & & \downarrow{\scriptstyle g[\epsilon]} \\ S & \xrightarrow[\iota]{} & S[\epsilon] \end{array}$$

$$\begin{array}{ccc} R[\epsilon] & \xrightarrow{\pi} & R \\ {\scriptstyle g[\epsilon]}\downarrow & & \downarrow{\scriptstyle g} \\ S[\epsilon] & \xrightarrow[\pi]{} & S \end{array}$$

Now let D be a derivation of R. It is straightforward to check that the function $f_D : R \to R[\epsilon]$ by $a \mapsto a + D(a)\epsilon$ is (1) a ring homomorphism and (2) satisfies $\pi \circ f_D = \mathrm{id}_R$. When D is the zero derivation, f_0 is the R algebra map $R \to R[\epsilon]$ previously noted. Conversely, suppose $f : R \to R[\epsilon]$ is a ring homomorphism that satisfies $\pi \circ f = \mathrm{id}_R$. Using this composition equality and the bijection $R^2 \to R + R\epsilon$, we can write

$f(a) = a + D_f(a)\epsilon$ for some function $D_f : R \to R$. Then the fact that f is a ring homomorphism easily implies that D_f is a derivation. It is also easy to check that

$$D_{(f_D)} = D \text{ and } f_{(D_f)} = f.$$

Thus we have a bijection between derivations of R and ring homomorphisms from R to the dual numbers over R whose composition with π is the identity.

We record this for future reference:

Proposition 1.1 *Let $R[\epsilon] = R + R\epsilon$ denote the ring of dual numbers over R, and let $\iota : R \to R[\epsilon]$ and $\pi : R[\epsilon] \to R$ be given by $a \mapsto a + 0\epsilon$ and $a + b\epsilon \mapsto a$, respectively. Then*

1. *If D is a derivation of R then $f_D : R \to R[\epsilon]$ by $f_D(a) = a + D(a)\epsilon$ is a ring homomorphism such that $\pi \circ f_D = id_R$.*

2. *If $f : R \to R[\epsilon]$ is a ring homomorphism such that $\pi \circ f = id_R$ and $D_f : R \to R$ is given by $f(a) = a + D_f(a)\epsilon$ then D is a derivation of R.*

3. *$D_{(f_D)} = D$ and $f_{(D_f)} = f$.*

The dual numbers construction also applies to derivations from a ring into another ring. For later reference, we record this as a proposition:

Proposition 1.2 *If $h : R \to S$ is ring homomorphism and $\delta : R \to S$ is an h-derivation of R into S then*

1. *$f_\delta : R \to S[\epsilon]$ by $a \mapsto h(a) + \delta(a)\epsilon$ is a ring homomorphism, and*

2. *if $f : R \to S[\epsilon]$ is a ring homomorphism such that $\pi \circ f = h$ then $\delta_f : R \to S$ given by $f(a) = h(a) + \delta_f(a)\epsilon$ is an h-derivation of R into S.*

Proposition 1.1 explains derivations in terms of ring homomorphisms. We show how to use this to extend derivations to ring extensions which are characterized by their ring homomorphism properties; in particular, to polynomial rings and rings of quotients.

We recall that a polynomial ring over the ring R in the set of indeterminants X is an R algebra T containing X with the universal property that for any R algebra U and set map $\phi : X \to U$ there is a unique R algebra homomorphism $g : T \to U$ with $g(x) = \phi(x)$ for all $x \in X$. As usual, we may assume $R \subset T$. Suppose that R is a differential ring. We have the ring homomorphism $R \to R[\epsilon] \to T[\epsilon]$ given by composing f_D with the inclusion of $R[\epsilon]$ to $T[\epsilon]$ from the inclusion of R in T making $T[\epsilon]$ an R algebra. Let $\phi_0 : X \to T$ be any map. Let $\phi : X \to T[\epsilon]$ be given by $x \mapsto x + \phi_0(x)\epsilon$. As $T[\epsilon]$ is an R algebra and $\phi : X \to T[\epsilon]$ is a set map, the universal property of T gives a unique ring homomorphism $f : T \to T[\epsilon]$ with $f(x) = \phi(x) = x + \phi_0(x)\epsilon$. Consider the composite $\pi \circ f$. For x in X, this composite is $x \mapsto x + \phi_0(x)\epsilon \mapsto x$, i.e., the inclusion $X \to T$. The identity homomorphism $T \to T$ is an R algebra homomorphism agreeing with $\pi \circ f$ on X, so it must be the unique such R algebra homomorphism. Thus $\pi \circ f$ is the identity. It follows from Proposition 1.1 that D_f defined by $f(t) = t + D_f(t)\epsilon$ is a derivation of T, and since $f(x) = x + \phi_0(x)\epsilon$ for $x \in X$, that $D_f(x) = \phi_0(x)$ for $x \in X$. Since f is an R algebra homomorphism, $f(r) = r + D(r)\epsilon$, so $D_f(r) = D(r)$ for $r \in R$. Thus we can extend set maps of X to T to derivations of T such that T is a differential extension of R. We record this as a proposition:

Proposition 1.3 *Let T be the polynomial ring on the set X over the differential ring R. Let $\phi : X \to T$ be any set map. Then there is a unique derivation D of T such that*

1. $D(x) = \phi(x)$ *for all* $x \in X$.

2. $D(r) = D_R(r)$ *for all* $r \in R$.

There is a special case of Proposition 1.3 which we will need later, so we record it now as a corollary. For ease of reference, we include the notation in the statement.

Corollary 1.1 *Let n be a positive integer and let T be the polynomial ring over R in the n^2 indeterminates x_{ij}, $1 \leq i \leq n$, $1 \leq j \leq n$. Let $X = [x_{ij}]$ be the $n \times n$ matrix with the indeterminates as entries. Let $A = [a_{ij}]$ be in $M_n(R)$. Then there is a unique derivation D on T extending D_R such that*

$$D(X) = AX.$$

That is $D(x_{ij}) = \sum_k a_{ik} x_{kj}$.

The same reasoning that establishes Proposition 1.3 establishes a similar result for derivations into.

Proposition 1.4 *Let T be the polynomial ring on the set X over the ring R. Suppose that $R \to S$ is a ring homomorphism and that $\delta : R \to S$ is a derivation of R into S. Let $\phi : X \to S$ be any set map. Then there is a unique derivation Δ of T into S such that*

1. $\Delta(x) = \phi(x)$ *for all* $x \in X$.

2. $\Delta(r) = \delta(r)$ *for all* $r \in R$.

As an application of Proposition 1.3, consider a differential ring homomorphism $f : R \to S$. Let $\{a_i \in S | i \in I\}$ be a set of generators of S as an R algebra, let $X = \{x_i | i \in I\}$ be a set of indeterminates, let T be the polynomial ring in X over R, and let $F : T \to S$ be given by $F(x_i) = a_i$ for all $i \in I$. By construction, F is surjective. For all $i \in I$, suppose $D_S(a_i) = b_i$, and that $y_i \in T$ is such that $F(y_i) = b_i$. By Proposition 1.3, there is a derivation D_T with $D_T(x_i) = y_i$ for all $i \in I$. Then $F(D_T(x_i)) = D_S(F(x_i))$ for all $i \in I$ and it follows that F is a differential ring homomorphism.

When X in Proposition 1.3 is the singleton $\{x\}$, so $T = R[x]$, and $\phi(x) = g \in R[x]$, we compute $D(\sum_{i=0}^n r_i x^i)$. For the term $r_i x^i$, $i \neq 0$, we have $D(r_i x^i) = D(r_i)x^i + r_i i x^{i-1} g$ and for $i = 0$ we have $D(r_0)$. Thus

$$D(\sum_{i=0}^n r_i x^i) = \sum_{i=0}^n D(r_i)x^i + (\sum_{i=1}^n i r_i x^{i-1})g.$$

Suppose $f(x)$ denotes $\sum_{i=0}^n r_i x^i$. In the above formula, the first summand on the right-hand side is the result of applying D to the coefficients of f; this is usually denoted $f^D(x)$, and the coefficient of g in the second summand is the derivative of f when $D(x) = 1$, which is usually denoted $f'(x)$. Thus the formula can also be written

$$D(f(x)) = f^D(x) + f'(x)g \ (*) \ .$$

We want to apply this to a ring of the form $S = R[x]/(f)$ where R is a differential ring, and we want to find a derivation of S such that

$R \to S$ is a differential homomorphism. That is, we want to find an R algebra homomorphism $S \to S[\epsilon]$. To do so, we look for an R algebra homomorphism $R[x] \to R[x][\epsilon]$ such that the composite $R[x] \to R[x][\epsilon] \to S[\epsilon]$ sends f to zero. The first map in this composite sends f to $f + D(f)\epsilon$; for this to go to zero in $S[\epsilon]$, we need $D(f) \in (f)$. Now suppose D is given by $D(x) = g \in R[x]$. By the above formula, $D(f) \in (f)$ if and only if there is an $h \in R[x]$ such that $f^D + f'g = hf$ (**).

Sometimes this happens: for instance, for any differential ring R, if $f = x - a$ for some $a \in R$, we can take $g = D(a)$, and then $h = 0$ works. On the other hand, if R is the polynomial ring $\mathbb{Z}[t]$ with $D(t) = 1$ and $f = x^2 + t$, then g and h have to satisfy $1 + 2xg = h(x^2 + t)$. Identify $R[x]$ with $\mathbb{Z}[t, x]$. If we set $t = 0$, the equation implies that $1 + 2xg(0, x) = h(0, x)(x^2)$ in $\mathbb{Z}[x]$, which is impossible. This provides an example of an injective ring extension $R = \mathbb{Z}[t] \to \mathbb{Z}[t][x]/(x^2 + t) = S$ and a derivation of R that can't be extended to S.

If both $f^D = 0$ and $f' = 0$ then for any g (**) holds with $h = 0$. For example, if $R = \mathbb{Z}/3\mathbb{Z}$ (so $D_R = 0$) and $f = x^3$ then $f^D = D(1)x^3 = 0$ and $f' = 3x^2 = 0$. Then, taking $g = x + 1$, we have a derivation of $S = R[x]/(x^3)$. Let $\overline{x}$ be the image of x in S. Then $\overline{x}$ is nilpotent ($\overline{x}^3 = 0$) but $D_S(\overline{x}) = \overline{x} + 1$ is not nilpotent.

As another application, we consider the case where R is a field (a differential ring which is a field is called a *differential field*) and the polynomial $f(x) \in R[x]$ is irreducible (of degree at least one, of course). Since R is a field, we may assume f is monic. We further assume that f and f' are relatively prime (which will always be the case if R has characteristic zero). Then there are $s, t \in R[x]$ such that $sf + tf' = 1$. Using the Euclidean Algorithm, we can write $t = qf + r$ where r is of smaller degree than f so that $1 = rf' + (s + qf')f$. Note that $r \neq 0$ since f is of positive degree, and hence that r maps to a non-zero element of $R[x]/(f)$. Let $g = -f^D r$. Then $f^D + f'g = f^D - f'rf^D = f^D(1 - rf') = f^D(s + qf')f = hf$ where $h = f^D(s + qf')$. By formula (**), this means that $D(x) = g$ defines a derivation of $R[x]$ which induces a derivation of $S = R[x]/(f)$ such that $R \to S$ is a differential homomorphism. Note that since f is irreducible, S is also a (differential) field; regarding $R \to S$ as the inclusion, we say that S is a *differential field extension* of R. In fact, we call any differential homomorphism between differential fields a differential field extension.

We can ask if there are any other choices for g in this application. Suppose we have some g_0 and h_0 such that $f^D + f'g_0 = h_0 f$. Then $(h_0 - h)f = (f^D + f'g_0) - (f^D + f'g) = (g_0 - g)f'$. Since f and f' are relatively prime, this means that f is a factor of $g_0 - g$ so that $g_0 = g + pf$ for some $p \in R[x]$. Conversely, if we take any $p \in R[x]$ and define g_0 to be $g + pf$. Then $f^D + f'g_0 = f^D + f'g + f'pf = (h + f'p)f = h_0 f$ where h_0 is defined to be $h + f'p$. Thus g is unique modulo f.

Let $\overline{x}$ be the image of x in the field S. The calculation just made shows that $g(\overline{x})$, namely $D_S(\overline{x})$, is uniquely determined: there is only one derivation of S which extends the derivation on R.

For later reference, we record this result:

Proposition 1.5 *Let R be a differential field of characteristic zero, let $S \supset R$ be a field extension, and let $\alpha \in S - R$ be algebraic over R with monic minimal polynomial $f \in R[x]$. Then there is a unique extension of D_R to $R(\alpha)$ given by*

$$D(\alpha) = \frac{f^{D_R}(\alpha)}{f'(\alpha)}.$$

Next, we look at extensions of derivations to rings of quotients. Let R be a differential ring and let $\mathcal{S}$ be a multiplicatively closed subset of R not containing 0. Let $S = \mathcal{S}^{-1}R$. We seek a derivation on S making the natural map $R \to S$ by $r \mapsto r/1$ a differential homomorphism; that is, we seek an R algebra homomorphism $S \to S[\epsilon]$. By the universal property of rings of quotients, any homomorphism $S \to S[\epsilon]$ is determined uniquely by the homomorphism $R \to S \to S[\epsilon]$, and exists provided this homomorphism sends elements of $\mathcal{S}$ to units. Consider the map $R \to R[\epsilon] \to S[\epsilon]$ where the first map comes from D_R and the second from $R \to S$. Under the composite, $x \in \mathcal{S}$ maps to $x/1 + D(x)/1\epsilon$, and since $(x/1 + D(x)/1\epsilon)(1/x - D(x)/x^2\epsilon) = 1$, the composite maps x to a unit, and hence extends to a homomorphism $S \to S[\epsilon]$. This latter sends $1/x$ to $(x/1 + d(x)/1\epsilon)^{-1}$, which we just saw is $1/x - D(x)/x^2\epsilon$. Thus the homomorphism sends $a/x = a/1 \cdot 1/x$ to $(a/1 + D(a)/1\epsilon)(1/x - D(x)/x^2\epsilon) = a/x + (D(a)/x - aD(x)/x^2)\epsilon = a/x + ((D(a)x - aD(x))/x^2)\epsilon$. By Proposition 1.1, the homomorphism $S \to S[\epsilon]$ thus corresponds to a derivation of S with $D_S(a/x) = (D_R(a)x - aD_R(x))/x^2$, which also implies that $R \to S$ is a differential homomorphism. We record this:

Proposition 1.6 *Let R be a differential ring, and let $\mathcal{S}$ be a multiplicatively closed subset of R not containing 0. Then $S = \mathcal{S}^{-1}R$ is a differential ring with derivation $D_S(a/x) = (D_R(a)x - aD_R(x))/x^2$ and $R \to S$ is a differential homomorphism.*

If R happens to be a differential domain then Proposition 1.6 implies that the quotient field of R is a differential field extension of R.

1.3 FIELDS

Suppose that F is a differential field and that $E \supset F$ is an extension field. From field theory, we know that E is an algebraic extension of a transcendental subextension $F(X)$ of F. By Proposition 1.3 we know that D_F can be extended to one of $F[X]$ and by Proposition 1.6 that derivation can be extended to a derivation $D_{F(X)}$ of $F(X)$.

Thus $E \supset F(X)$ is an algebraic extension of the differential field $F(X)$. Now we suppose that F, and hence $F(X)$, is of characteristic zero. Then using Proposition 1.5 we can extend the derivation to E:

Lemma 1.1 *Let F be a differential field of characteristic zero. Let $E \supset F$ be an algebraic extension. Then there is a unique derivation D_E of E extending D_F.*

Proof. First we tackle existence. By Zorn's lemma we can find a maximal subextension E_0, $E \supseteq E_0 \supseteq F$, such that there is a derivation D of E_0 extending D_F. If $E \supsetneq E_0$, let $\alpha \in E - E_0$. Of course, α is algebraic over E_0. By Proposition 1.5 D can be extended to $E_0(\alpha)$, which then is a strictly larger subextension of E over F that E_0 with a derivation extending extending D_F. This contradiction to the maximality of E_0 proves that $E = E_0$ has a derivation D extending D_F.

It remains to show that D is unique. Let $\alpha \in E$ and let $f \in F[x]$ be the unique monic irreducible polynomial of α over F. Then differentiating $f(\alpha) = 0$ gives $f^{D_F}(\alpha) + f'(\alpha)D(\alpha) = 0$. Since the characteristic is zero, $f'(\alpha) \neq 0$, so $D(\alpha) = -f^{D_F}(\alpha)/f'(\alpha)$. Since the right-hand side depends only on α and D_F D is uniquely determined by D_F.

Combining Lemma 1.1 and the remarks preceding about extending derivations to transcendental extension we then have

Corollary 1.2 *Let F be a differential field of characteristic zero and let $E \supset F$ be a field extension. Then there is a derivation D_E of E extending D_F*

We can apply these results to show the existence of many kinds of derivations.

Remark 1.1 *Let F be a field of characteristic zero and assume $a \in F$ is not algebraic over $\mathbb{Q}$. Then there is a derivation of F such that $D(a) = 1$. By assumption, the subfield $\mathbb{Q}(a)$ is purely transcendental extension of $\mathbb{Q}$ so it has a derivation with the desired property, and then by the Corollary this derivation can be extended from $\mathbb{Q}(a)$ to F.*

Remark 1.2 *Let F be a field of characteristic zero and assume $a, b \in F$ are not algebraically dependent over $\mathbb{Q}$. Then there is a derivation D of F such that $D(b) = 0$ and $D(a) = ba$. For by assumption the subfield $\mathbb{Q}(a,b)$ is a pure transcendental extension of $\mathbb{Q}$ and hence it has a derivation with the desired properties. Then by Corollary 1.2 that derivation extends from $\mathbb{Q}(a,b)$ to F. Note that b is a constant of D and a behaves like the function e^{bx}.*

Remark 1.3 *Let F be a differential field of characteristic zero, let $E \supset F$ be an extension field and let X be a transcendence base of E over F. Then E has at least $|X^X|$ derivations extending the derivation of F. For any map $X \to X$ gives rise to a derivation of the subfield $F(X)$ of E, and since $E \supset F(X)$ is algebraic all these derivations extend to E.*

Derivations of E do not have to preserve X of course; any map $X \to E$ also gives rise to a derivation.

If we apply this remark to $\mathbb{R}$ or $\mathbb{C}$ as extensions of $\mathbb{Q}$, the latter with 0 derivation, we see that $\mathbb{R}$ and $\mathbb{C}$ have uncountably many derivations. There exist derivations such that $D(\pi) = 0$, or $D(\pi) = 1$, or $D(\pi) = e\pi$. There is no constructive way to compute any of these derivations.

1.4 DIFFERENTIAL IDEALS

Suppose $f : R \to S$ is a differential homomorphism of differential rings. Let $a \in R$ and suppose $f(a) = 0$. Then $0 = D_S(0) = D_s(f(a)) = f(D_R(a))$. In other words, if I is the kernel of f then D_R carries I to itself. Any such ideal of R stable under D is called a *differential ideal*. Note that $\{0\}$ is a differential ideal. Generally speaking, and following the usual conventions of ring theory, we do not call R an ideal of R. If this point ineeds emphasis, we call an ideal a *proper* ideal to indicate it is not R. If J is a differential ideal of R, and S denotes the ring R/J, then the abelian group endomorphism D_R of R induces an abelian group endomorphism $a + J \mapsto D_R(a) + J$ of S. We denote this endomorphism D_S. The Liebnitz rule for D_R shows that D_S is a derivation of S and that $R \to S$ by $a \mapsto a + J$ is a differential homomorphism (with kernel J, of course). If I is a differential ideal of R which cotains J then the ideal I/J of R/J is a differential ideal. Conversely, the inverse image under $R \to R/J$ of a differential ideal of R/J is a differential ideal of R which contains J. In fact for any differential ring homomorphism with source R then the inverse image of a differential ideal of the target is a differential ideal of R containing the kernel of the differential homomorphism.

It is straightforward to check that if I and J are differential ideals, so are $I \cap J$, $I + J$, and IJ. A differential ideal could be maximal, or prime, or nilpotent; in such cases, we say we have a differential maximal ideal, or a differential prime ideal, or a differential nilpotent ideal. On the other hand, we could also have a maximal differential ideal: this would be a differential ideal I (different from R, of course) such that any strictly larger differential ideal must be R itself. In other words, I is maximal among differential ideals with respect to not containing 1. Expressed this way, and using the observation that the union of a chain of differential ideals is a differential ideal, a Zorn's Lemma argument shows that any differential ideal is contained in a maximal differential ideal. Since we could start with the zero ideal, we know that maximal differential ideals exist. There is no reason that a maximal differential ideal needs to be a maximal ideal.

For example, consider the differential ring $R = \mathbb{Q}[x]$ with derivation $D_R(f) = f'$. Suppose I is a non-zero differential ideal of R, different from R. Then I is principal, say generated by f of degree $n \geq 1$. Since I is differential, $f' \in I$, which means that f is a factor of f'. Because f' has degree $n-1 > 0$, this can't happen. Thus for R zero is a maximal differential ideal, but it is not a maximal ideal of R. It is a prime ideal of R, and this is quite general, at least for rings which, like R, are $\mathbb{Q}$ algebras.

To begin the proof of this fact, we begin with a lemma:

Lemma 1.2 *Let R be a differential ring, $a \in R$, and m, p, q nonnegative integers with m and $p + q$ positive. Suppose $ma^p D(a)^q = 0$. Then there are positive integers n and r such that $nD(a)^r = 0$.*

Proof. If $p = 0$ the conclusion holds with $n = m$ and $r = p$. We argue by contradiction. If the conclusion is false, suppose the triple (m,p,q) is such that p is minimal. Of course, p is positive. If $q = 0$ so $0 = ma^p$then differentiating gives $0 = mpa^{p-1}D(a)$, and the triple $(mp,p-1,1)$ has a second entry less than p, contrary to our choice. Otherwise, differentiating $0 = ma^pD(a)^q$ gives $0 = mpa^{p-1}D(a)^{q+1} + m(q-1)a^pD(a)^{q-1}D^2(a)$; if we multiply this equation by $D(a)$ the second summand contains the factor $ma^pD(a)^q$ which is zero, and hence $0 = m(p-1)a^{p-1}D(a)^{q+1}$ and again the triple $(m(p-1),p-1,q+1)$ is contrary to our choice. Hence the conclusion is true.

As a special case of Lemma 1.2, we can consider the derivative of a nilpotent element:

Corollary 1.3 *Let R be a differential ring and $a \in R$ a nilpotent element. Then there are positive integers n and r such that $nD(a)^r = 0$. In particular, if R has no $\mathbb{Z}$ torsion then $D(a)$ is nilpotent. Hence, in the case R has no $\mathbb{Z}$ torsion, the ideal of nilpotent elements of R is differential.*

Proof. Let p be a positive integer such that $a^p = 0$. Then $1 \cdot a^pD(a)^0 = 0$, so by Lemma 1.2 there are positive integers n and r such that $nD(a)^r = 0$. If R is $\mathbb{Z}$ torsion-free, this means $D(a)^r = 0$ so $D(a)$ is nilpotent.

Lemma 1.2 also applies to zero divisors. Suppose a and b are nonzero elements of the differential ring R and that $0 = ab$. Differentiate this equation: then $0 = D(a)b + aD(b)$. Multiply this equation by a; the first summand vanishes (since $ab = 0$) and we are left with $0 = a^2D(b)$. This is the case $i = 1$ of the formula $0 = a^{i+1}D^i(b)$ (and $0 = ab$ is the case for $i = 0$). A straightforward induction shows that it holds for all i:

Lemma 1.3 *Let R be a differential ring and $a, b \in R$. Suppose that $ab = 0$. Then $a^{n+1}D^n(b) = 0$ for all n.*

Proof. As we have already noted, $ab = 0$ is the case $n = 0$ of the conclusion. Suppose that for some i $a^{i+1}D^i(b) = 0$ (the case $n = i$ of the conclusion). Differentiate this equation and multiply by a. This yields $a(i+1)a^iD(a)D^i(b) + aa^{i+1}D^{i+1}(b) = 0$. The first summand is a multiple of $a^{i+1}D^i(b)$, which is assumed to be 0, so the second summand $a^{i+2}D^{i+1}(b)$ is zero as well; this is the case $n = i+1$ of the conclusion. Hence the conclusion follows for all n by induction.

We apply Lemma 1.3 in the case where b is not in any differential ideal.

Corollary 1.4 *Let R be a differential ring and $a, b \in R$. Suppose that $ab = 0$ and that b is not contained in any differential ideal. Then a is nilpotent. In particular, if $\{0\}$ is a maximal differential ideal of R then zero divisors are nilpotent.*

Proof. Since b is not in any differential ideal, the ideal generated by b and its derivatives is R. That is, there is an n and elements $r_0, \ldots, r_n \in R$ such that $1 = \sum r_i D^i(b)$. Thus $a^{n+1} = \sum r_i a^{n+1} D^i(b)$, and by Lemma 1.3 $a^{n+1}D^i(b) = 0$ for $i \leq n$. Thus $a^{n+1} = 0$. If $\{0\}$ is a maximal differential ideal and $ab = 0$ with $a \neq 0$ and $b \neq 0$ then neither a nor b is contained in a differential ideal, so by what has been shown, both are nilpotent.

Note that Lemma 1.3 and hence Corollary 1.4 make no assumptions about $\mathbb{Z}$ torsion. If there is no $\mathbb{Z}$ torsion in R, then we can further conclude that R has no zero divisors.

Proposition 1.7 *Suppose R is a differential ring with no $\mathbb{Z}$ torsion. Suppose that $\{0\}$ is a maximal differential ideal of R. Then R is an integral domain.*

Proof. By Corollary 1.3 the ideal of nilpotent elements of R is differential, hence must be $\{0\}$. By Corollary 1.4 the zero divisors of R are nilpotent.

If I is a maximal differential ideal of R then $\{0\}$ is a maximal differential ideal of R/I. To apply Proposition 1.7 to R/I, we need to know that R/I has no $\mathbb{Z}$ torsion. This may not be true even if R has no $\mathbb{Z}$ torsion. One easy way to guarantee that no $\mathbb{Z}$ torsion passes to quotients is to require that R be a $\mathbb{Q}$ algebra.

Corollary 1.5 *Let R be a differential ring which is a $\mathbb{Q}$ algebra. Then the maximal differential ideals of R are prime ideals of R. Moreover, the radical of a differential ideal of R is also a differential ideal.*

Proof. Let I be a maximal differential ideal of R. Then by Proposition 1.7, R/I is an integral domain, so I is a prime ideal of R. Let J be a differential ideal of R. Then by Corollary 1.3, the set of nilpotent elements of R/J is a differential ideal. The inverse image of this set in R is the radical of J, which thus is also a differential ideal.

1.5 DIFFERENTIAL MODULES

Let R be a differential ring. A *differential R-module* is a pair (M, D_M) consisting of an R module M and a designated additive endomorphism D_M satisfying $D_M(rm) = D_R(r)m + rD_M(m)$ for all $r \in R$ and $m \in M$.

When the context is clear, we refer to the differential module by its first entry. For example R, with the endomorphism D_R, is a differential R module. If I is a differential ideal of R, then I is a differential R module (with the endomorphism $D_I = D_R|I$), and R/I is a differential R module (with the endomorphism $D_{R/I}(r+I) = D_R(r)+I$). If M and N are differential R modules then an R module homomorphism $f : M \to M$ is *differential* if $D_N \circ f = f \circ D_M$. The identity $R \to R$ is a differential R module homomorphism, as is, for a differential ideal I of R, the inclusion $I \to R$ and the projection $R \to R/I$.

The kernel of D_M is denoted M^D. It is a module over the ring R^D. Elements of M^D are called constants (of M).

As we saw above and recorded in Proposition 1.1, derivations of R correspond to certain ring homomorphisms $R \to R[\epsilon]$, the derivation $D = D_R$ of R corresponding to the homomorphism $f_D : R \to R[\epsilon]$ being $f_D(r) = r + D(r)\epsilon$. Let M be an R module, and let $M[\epsilon]$ denote $R[\epsilon] \otimes_R M = M \oplus M\epsilon$. By extension of scalars $M[\epsilon]$ is an $R[\epsilon]$ module. Since $R[\epsilon]$ is an R algebra via f_D, this means $M[\epsilon]$ is also an R module via f_D. (It is also an R module using any other derivation on R. To conserve notation, we will restrict attention to the designated derivation D_R.) The action on $M[\epsilon]$ via f_D is as follows:

$$r \star (a+b\epsilon) = (r+D(r)\epsilon)(a+b\epsilon) = ra + (rb + D(r)a)\epsilon$$

If M is a differential module with derivation D_M, then we can define a function $F : M \to M[\epsilon]$ by $F(m) = m + D_M(m)\epsilon$. Then $F(rm) = rm + D_M(rm)\epsilon = rm + (rD_M(m) + D(r)m)\epsilon = r \star (m + D_M(m))\epsilon = r \star F(m)$. In other words, F is an R module homomorphism. This is analogous to the first assertion of Proposition 1.1, and in fact, we have a complete analogy to all the proposition.

Proposition 1.8 *Let M be an R module, Regard $M[\epsilon]$ as an R module via f_D where $D = D_R$ and let $\iota : M \to M[\epsilon]$ and $\pi : M[\epsilon] \to M$ be given by $a \mapsto a + 0\epsilon$ and $a + b\epsilon \mapsto a$ respectively. Then*

1. *If $D = D_M$ is a derivation of M, then $F_D : M \to M[\epsilon]$ by $F_D(a) = a + D(a)\epsilon$ is an R module homomorphism such that $\pi \circ F_D = id_M$.*

2. *If $F : M \to M[\epsilon]$ is an R module homomorphism such that $\pi \circ F = id_M$ and $D_F : M \to M$ is given by $F(a) = a + D_F(a)\epsilon$ then D is a derivation of M.*

3. *$D_{(F_D)} = D$ and $F_{(D_F)} = F$*

Proof. For the first assertion, we have already shown that F_D is an R module homomorphism; that $\pi \circ F_D = \mathrm{id}_M$ is clear. For the second assertion, suppose we have such an R module homomorphism F so that $F(m) = m + D_F(m)\epsilon$. Being an R module homomorphism says that $F(rm) = rm + D_F(rm)\epsilon$ equals $r \star F(m) = (r + D(r)\epsilon)(m + D_F(m)\epsilon) = (rm + (rD_F(m) + D(r)m)\epsilon)$. Comparing coefficients of ϵ shows that $D_F(rm) = rD_F(m) + D(r)m$, and hence D_F is a derivation of M. We omit the verification of the third assertion.

Let M be a differential R module with derivation D_M. We are going to observe that D_M induces a derivation on the symmetric algebra $S_R(M)$ of M over R which makes the symmetric algebra a differential extension of R. We recall that $S_R(M)$ is the quotient of the tensor algebra

$T_R(M)$ by the two sided ideal generated by $\{m \otimes n - n \otimes m | \forall m, n \in M\}$ and that it is characterized by being universal with respect to commutative R algebras to which M maps as an R module. Both of these characterizations provide (separately) a proof that D_M extends to a ring derivation.

Proposition 1.9 *Let (M, D_M) be a differential R module. Then there is a derivation of $S_R(M)$ which coincides with D_M on M and D_R on R.*

Before starting the proof, we recall the construction of $S_R(M)$. For $n = 0, 1, 2, \ldots$ define $T_R^n(M)$ as follows: $T_R^0(M) = R$, $T_R^1(M) = M$, $T_R^n(M) = M \otimes_R M \otimes_R \cdots \otimes_R M$ (n factors) for $n \geq 2$. Then $T_R(M) = \bigoplus_{n \geq 0} T_R^n(M)$. Then $T_R(M)$ is a graded (non-commutative) R algebra, the ring multiplication given by tensor product: if $a \in T_R^n(M)$ and $b \in T_R^m(M)$ then $a \cdot b = a \otimes b \in T_R^{n+m}(M)$. Moreover, $R \subset T_R(M)$ as a subring via $R = T_R^0(M)$ and $M \subset T_R(M)$ as an R submodule via $M = T_R^1(M)$. Let I be the twosided ideal of $T_R(M)$ generated by $\{m \otimes n - n \otimes m | \forall m, n \in M\}$. Let $I^n = I \cap T_R^n(M)$. Then $S_R(M) = T_R(M)/I$ is commutative. Since I is generated by homogeneous elements of degree two, $I = \bigoplus_{n \geq 0} I^n$ and $S_R(M) = \bigoplus_{n \geq 0} S_R^n(M)$, where $S_R^n(M) = T_R^n(M)/I^n$. Sinc $I^0 = I^1 = 0$, we still have $R \subset S_R(M)$ and $M \subset S_R(M)$; denote this latter inclusion h. If T is any commutative R algebra and $f : M \to T$ is an R module homomorpihsm then f induces R module homomorphisms $T_R^n(M) \to T$ by $a_1 \otimes \cdots \otimes a_n \mapsto f(a_1) \cdot \cdots \cdot f(a_n)$ and the direct sum of these is a ring homomorphism $T_R(M) \to T$, which, since T is commutative factors through $S_R(M)$. The resulting R algebra homomorphism $S_R(M) \to T$ has the property that when preceded by h, it is the original R module homomorphism f and that it is uniquely determined by this.

Proof. (First proof of Proposition 1.9) The derivation D_M induces a derivation on each $T_R^n(M)$ and hence on their direct sum (as an R module). But it is clear from the definition of the multiplication on $T_R(M)$ that this is a derivation of $T_R(M)$ as an R algebra. Then we note that I is a differential ideal: $D(m \otimes n - n \otimes m) = D(m) \otimes n + m \otimes D(n) - D(n) \otimes m - n \otimes D(m) = (D(m) \otimes n - n \otimes D(m)) + (m \otimes D(n) - D(n) \otimes n) \in I$. Thus $S_R(M) = T_R(M)/I$ is a commutative differential ring whose derivation is D_R on R and D_M on M by construction.

Proof. (Second proof of Proposition 1.9) Consider the $R[\epsilon]$ algebra $S_R(M)[\epsilon] = R[\epsilon] \otimes_R S_R(M)$. We regard $S_R(M)[\epsilon]$ as an R algebra using $R \to R[\epsilon]$ by $r \mapsto r + D_R(r)\epsilon$. The map $M \to S_R(M)[\epsilon]$ by $m \mapsto m + D_M(m)\epsilon$ is then an R module homomorphism, which by the universal property produces an R algebra map $F : S_R(M) \to S_R(M)[\epsilon]$ such that $F(m) = m + D_M(m)$ for $m \in M$. As before, let $\pi : S_R(M)[\epsilon] \to S_R(M)$ be given by $\pi(\epsilon) = 0$. Then $\pi \circ F$ is a map $S_R(M) \to S_R(M)$ with $\pi \circ F(m) = m$ for $m \in M$. Since the identity $S_R(M) \to S_R(M)$ also has this property, the universal property says $\pi \circ F$ is the identity. Define $D : S_R(M) \to S_R(M)$ by $F(s) = s + D(s)\epsilon$. By Proposition 1.1, D is a derivation of $S_R(M)$, which then is observed to satisfy the restriction requirements.

A given R module has, in general, many possible differential structures:

Proposition 1.10 *Let M be an R module and let D_1 and D_2 be additive endomorphisms of M satisfying $D_i(rm) = D_R(r)m + rD_i(m)$ for all $r \in R$ for $i = 1,2$. Then $D_1 - D_2$ is an R module endomorphism of M. Conversely, suppose D_M is an additive endomorphism of M satisfying $D_M(rm) = D_R(r)m + rD_M(m)$ for all $r \in R$ and let T be an R module endomorphism of M. Then the additive endomorphism $D_3 = D_M + T$ of M satisfies $D_3(rm) = D_R(r)m + rD_3(m)$ for all $r \in R$.*

The proof of Proposition 1.10 is a straightforward calculation.

If M and N are differential R modules then $M \oplus N$ is a differential R module with $D_{M\oplus N} = D_M \oplus D_N$,$M \otimes_R N$ is a differential R module with $D_{M\otimes N} = D_M \otimes id_N + id_M \otimes D_N$. and $\mathrm{Hom}_R(M,N)$ is a differential R module with $D_{\mathrm{Hom}(M,N)}(T) = TD_M - D_N T$.

As a special case of this latter, we consider when $N = (R, D_R)$. Then we have a differential structure on $\mathrm{Hom}_R(M,R)$ given by $D(f)(m) = D_R(f(m)) - f(D_M(m))$; we call this the *differential dual* of M.

All the above constructions are functorial. We record this in the following proposition:

Proposition 1.11 *Let M, N, M_1, N_1, M_2, N_2 be differential R modules. Let $f : M_1 \to N_1$ and $g : M_2 \to N_2$ be differential homomorphisms. Then*

1. *$M_1 \oplus N \to M_2 \oplus N$ by $(a,b) \mapsto (f(a), b)$ and $M \oplus N_1 \to M \oplus N_2$ by $(a,b) \mapsto (a, g(b))$ are differential homomorphisms.*

2. *$M_1 \otimes N \to M_2 \otimes N$ by $a \otimes b \mapsto f(a) \otimes b$ and $M \otimes N_1 \to M \otimes N_2$ by $a \otimes b \mapsto a \otimes g(b))$ are differential homomorphisms.*

3. *$Hom_R(M_2,N) \to Hom_R(M_1,N)$ by $T \mapsto Tf$ is a differential homomorphism and $Hom_R(M,N_1) \to Hom_R(M,N_2)$ by $T \mapsto gT$ is a differential homomorphism.*

4. *$M \to Hom_R(Hom_R(M,N),N)$ by $m \mapsto (T \mapsto T(m))$ is a differential homomorphism.*

Proof. We omit the calculations for the first three assertions. For the last, we recall the notation $\hat{m}$ for the map $T \mapsto T(m)$, and we calculate $D(\hat{m})$. This latter is given by $D(\hat{m})(T) = D_N(\hat{m}(T)) - \hat{m}(D(T)) = D_N(T(m)) - D(T)(m)$. Since $D(T)(m) = D_N(T(m)) - T(D_M(m))$, we have $D(\hat{m})(T) = T(D_M(m)) = \widehat{D_M(m)}(T)$. This holds for all T so $D(\hat{m}) = \widehat{D(m)}$, proving the fourth assertion.

1.6 R LINEAR HOMOMORPHISMS OF DIFFERENTIAL MODULES

We are now going to consider some additional constructions with R modules which involve homomorphism modules, tensor products, and duals.

First some general comments. Category theory has taught us that the morphisms, even the isomorphisms, between objects matter. In some cases, such as Tannaka Duality theory for algebraic groups, it teaches us that the category of representations of a structure determines that structure. Morita Duality for rings may be viewed analogously. After all, the category of modules of a ring is the category of representations of the ring as endomorphisms of abelian groups. In the Morita Theory, the category of modules for a ring determines the ring up to tensoring with an endomorphism ring of a module; if we know the ring is commutative then the ring is determined. In the case of a differential commutative ring R such as we are considering here, although we could ask about a differential duality theory, it actually turns out to be more productive so simply consider R linear maps between differential R modules, In particular, if V and W are differential R modules, we are going to consider them along with the differential R module $\mathrm{Hom}_R(V,W)$ and the fundamental morphism

$$\mathrm{Hom}_R(V,W)\otimes V \to W$$

which relates all three.

Regarding duals, we will simplify by writing $(\cdot)^*$ for $\mathrm{Hom}_R(\cdot,R)$. Then, for double duals, we write $\hat{v}$ for the element of V^{**} given by $\hat{v}(f)=f(v)$. For the moment, we just consider two R modules V and W, which will later be taken to be differential R modules. To begin, we have the R module homomorphism $\mu:\mathrm{Hom}_R(V,W)\otimes_R V\to W$ given by $\mu(T\otimes v)=T(v)$. We call the dual of μ, $\gamma: W^*\to(\mathrm{Hom}_R(V,W)\otimes_R V)^*$, so $\gamma(\phi)(T\otimes v)=\phi(T(v))$.

Now suppose both V and W are finitely generated projective R modules. Then $(\mathrm{Hom}_R(V,W)\otimes_R V)^*$ can be identified with $\mathrm{Hom}_R(V,W)^*\otimes V^*$. Also $\mathrm{Hom}_R(V,W)$ can be identified with $V^*\otimes W$. We establish the following notation: if $\phi\in V^*$ and $w\in W$ then $S_{\phi,w}:V\to W$ is defined as $S_{\phi,w}(v)=\phi(v)w$; these symbols are bilinear and $V^*\otimes_R W\to \mathrm{Hom}_R(V,W)$ by $\phi\otimes w\mapsto S_{\phi,w}$ induces an isomorphism. (We retain this "S" notation when V and W are replaced by any finitely generated projective R modules, for example, each other.)

So $\mathrm{Hom}_R(V,W)^*$ can be identified with $(V^*\otimes W)^*$ and this latter with $V^{**}\otimes W^*$, then using the identification of V^{**} and V we have

$(\mathrm{Hom}_R(V,W)\otimes_R V)^*$ identified with $W^*\otimes V\otimes V^*$ (permuting the tensor factors). Under these identifications, which are all natural and, if V and W are differential also identifications of differential modules, we have $\gamma: W^* \to W^*\otimes V\otimes V^*$ such that $\gamma(\phi) = \sum \psi_i \otimes v_i \otimes f_i$ provided, for all T and v, that $\phi(T(v)) = \sum \psi(T(v_i))f_i(v)$.

We used above the identification of $\mathrm{Hom}_R(V,W)^*$ with $W^*\otimes V$. Since $W^*\otimes V$ is identified with $\mathrm{Hom}_R(W,V)$, then $\mathrm{Hom}_R(W,V)$ is identified with $\mathrm{Hom}_R(V,W)^*$, where we are thinking of the latter as $W^*\otimes V$. We establish some notation for this last: if $\phi \in W^*$ and $v \in V$ then $m_{\phi,v} \in \mathrm{Hom}_R(V,W)^*$ is defined as $m_{\phi,v}(T) = \phi(T(v))$; these symbols are bilinear and $W^*\otimes_R V \to \mathrm{Hom}_R(V,W)^*$ by $\phi\otimes v \mapsto m_{\phi,v}$ induces an isomorphism. (As above, we retain this "m " notation V and W are replaced by any finitely generated projective R modules, for example, each other.)

We have already remarked about the bilinearity of the symbols $S_{\phi,w}$ (because it comes from $\phi\otimes w$) and $m_{\phi,v}$ (because it comes from $\phi\otimes v$). In Lemma 1.4 we will show that these correspondences are differential. This then gives the following formulae for the derivatives of these symbols:

$$D(S_{\phi,w}) = S_{D(\phi),w} + S_{\phi,D(w)} \text{ and } D(m_{\phi,w}) = m_{D(\phi),w} + m_{\phi,D(w)}.$$

If $S \in \mathrm{Hom}_R(W,V)$ and $T \in \mathrm{Hom}_R(V,W)$, we can apply S to T as a functional; we denote this as $S\star T \in R$. Explicitly, if $S = \sum_{g\in X, v\in Y} S_{g,v}$ where $X \subset W^*$ and $Y \subset V$ are finite subsets, then $S\star T = \sum_{g\in X,v\in Y} m_{g,v}(T)$.

Viewed this way, we have $\gamma : W^* \to \mathrm{Hom}_R(W,V)\otimes V^*$, where if $\gamma(\phi) = \sum S_i \otimes f_i$ then for all $T \in \mathrm{Hom}_R(V,W)$ and $v \in V$ we have $\sum S_i \star T f_i(v) = \phi(T(v))$.

Now suppose that v_i, f_i, $1 \leq i \leq n$ is a projective coordinate system of V and w_j, g_j, $1 \leq j \leq m$ is a projective coordinate system of W. Then for $v \in V$, $v = \sum f_i(v)v_i$ and for $w \in W$ $w = \sum g_j(w)w_j$. Let $T \in \mathrm{Hom}_R(W, V)$. Then $T(w) = \sum_i f_i(T(w))v_i = \sum_i f_i(\sum_j g_j(w)T(w_j))v_i = \sum_{i,j} f_i(T(w_j))g_j(w)v_i$. In our notations $f_i(T(w_j)) = m_{f_i,w_j}(T) \in \mathrm{Hom}_R(W, V)^*$ and $g_j(w)v_i = S_{g_j,v_i}(w)$. So

$$T = \sum_{i,j} m_{f_i,w_j}(T)S_{g_j,v_i}.$$

This shows that S_{g_j,v_i}, m_{f_i,w_j}, $1 \leq i \leq n$, $1 \leq j \leq m$ is a projective coordinate system for $\mathrm{Hom}_R(W, V)$. If both V and W are free, and the v_i and w_j are bases with dual bases f_i and g_j, then S_{g_j,v_i} is a basis of $\mathrm{Hom}_R(W, V)$ with dual basis m_{f_i,w_j}.

For any T, $T(v) = \sum T(v_i)f_i(v)$ so for $\phi \in W^*$ we have $\phi(T(v)) = \sum \phi(T(v_i))f_i(v)$. Thus with $\gamma : W^* \to W^* \otimes V \otimes V^*$, $\gamma(\phi) = \sum \phi \otimes v_i \otimes f_i = \phi \otimes \sum v_i \otimes f_i$.

The derivations D_V and D_W are determined by their values on the v_i and w_i: if $v \in V$, then since $v = \sum_i f_i(v)w_i$, $D(v) = \sum_i (D(f_i(v))v_i + f_i(v)D(v_i))$ and similarly for W. As $D(v_i) \in V$, we have $D(v_i) = \sum_j f_j(D(v_i))v_j$. For notational convenience, we temporarily denote $f_j(D(v_i))$ by a_{ji}, so $D(e_i) = \sum_j a_{ji}v_i$. Similarly, we let b_{ij} denote $g_j(D(w_i))$ so $D(w_i) = \sum_j b_{ji}w_j$. Note that the $n \times n$ matrix $A = [a_{ij}]$ (respectively the $m \times m$ matrix $B = [b_{ij}]$) determines D_V (respectively D_W). But not every $n \times n$ ($m \times m$) matrix determines a derivation of V (W); we will explore later which ones do.

A (respectively B) also determines the derivation on V^* (W^*). For $v \in V$ and $f \in V^*$, $f(v) = f(\sum_i f_i(v)v_i) = (\sum_i f(v_i)f_i)(v)$. Using $\hat{v}$ for the functional on V^* given by evaluation at v, this shows that $f_i, \hat{v}_i$ is a projective coordinate system on V^*. Hence, for $f \in V^*$, we have, as before, D on V^* determined by the $D(f_i)$, and that $D(f_i) = \sum_j \hat{v_j}(D(f_i))f_j$. For any $f \in V^*$, and $v \in V$, $D(f)(v) = D_R(f(v)) - f(D_V(v))$. Thus $\hat{v_j}(D(f_i)) = D(f_i)(v_j) = D_R(f_i(v_j)) - f_i(D_V(v_j))$. Temporarily let $c_{ij} = f_i(v_j)$. Then $D(f_i) = \sum_j (c'_{ij} - a_{ij})f_j$. Similarly, if $d_{ij} = g_i(w_j)$ the $D(g_i) = \sum_j (d'_{ij} - b_{ij})g_j$. If the v_i are a basis of V and the f_i are the dual basis the $f_i(v_j)$ is either 0 or 1, so $c'_{ij} = 0$. A similar remark applies to W.

Now we use these formulas to compute the derivatives of the projective coordinate system S_{g_j,v_i}, m_{f_i,w_j}, $1 \le i \le n$, $1 \le j \le m$ i for $\mathrm{Hom}_R(W,V)$.

Let $x_{ij} = S_{g_j,v_i}$ and $t_{ij} = m_{f_i,w_j}$. Above we computed that $D(g_j) = \sum_p (d'_{jp} - b_{jp}g_p)$ and $D(v_i) = \sum_q a_{qi}v_q$, so using the derivative formula above for x_{ij} and bilinearly we get

$$D(x_{ij}) = \sum_{p,q} ((d'_{jp} - b_{jp}) + a_{qi})x_{pq}.$$

The formulas for the derivatives of f_i and w_j then give

$$D(t_{ij}) = \sum_{p,q} ((c'_{ip} - a_{ip}) + b_{qj})t_{ij}.$$

These expressions can be applied to calculate the derivation on the symmetric algebra $S_R(\mathrm{Hom}_R(W,V))$.

And with $\gamma : W^* \to \mathrm{Hom}_R(W,V) \otimes V^*$ this means that $\gamma(\phi) = \sum S_{\phi,v_i} \otimes f_i$ since $(\sum S_{\phi,v_i} \otimes f_i)(T \otimes v)$ which equals $\sum m_{\phi,v_i}(T) f_i(v) = \sum \phi(T(v_i)) f_i(v) = \phi(T(v))$.

We have asserted that in these formulations γ is differential, which means, to take the first, that $D(\phi) \otimes \sum v_i \otimes f_i = D(\phi \otimes \sum v_i \otimes f_i)$ Since the right side is $D(\phi)\otimes(\sum v_i\otimes f_i)+\phi\otimes D(\sum v_i\otimes f_i)$, for γ to be differential we need $D(\sum v_i\otimes f_i)=0$. Under the identification of $V^*\otimes V$ with $\mathrm{Hom}_R(V,V)$ $f\otimes v$ becomes the endomorphism $x\mapsto f(x)v$, which means that $\sum v_i\otimes f_i$ is the endomorphism $x\mapsto\sum f_i(x)v_i$ and since the v_i, f_i are a projective coordinate system $\sum f_i(x)v_i=x$. That is, $\sum v_i\otimes f_i\mapsto I$. And because the map is an isomorphism, this shows that $\sum v_i\otimes f_i$ is the same element of the tensor product regardless of which projective coordinate system is used. Since for the identity I we have $D(I)=0$, we also have $D(\sum v_i\otimes f_i)=0$ provided the identification of $V^*\otimes V$ with $\mathrm{Hom}_R(V,V)$ is differential as asserted. For completeness, we verify this in more generality.

Lemma 1.4 *Let P and Q be differential R modules which are finitely generated and projective as R modules. Then $P^*\otimes Q\to Hom_R(P,Q)$ by $f\otimes q\mapsto(x\mapsto f(x)q)$ is differential.*

Proof. Let $f\in P^*$ and $q\in Q$. Let T denote the homomorphism $x\mapsto f(x)q$ so that T is the image of $f\otimes q$. Since $D(f\otimes q)=D(f)\otimes q+f\otimes D(q)$, the right-hand side is sent to the homomorphism $x\mapsto D(f)(x)q+f(x)D(q)$. Thus the lemma is asserting that $D(T)(x)=D(f)(x)q+f(x)D(q)$. By definition, $D(T)(x)=D(T(x))-T(D(x))=D(f(x)q)-f(D(x))q$. Since $D(f(x)q)=D(f(x))q+f(x)D(q)$, we have $D(T)(x)=D(f(x))q+f(x)D(q)-f(D(x))q=(D(f(x))-f(D(x))q+f(x)D(q)$. Since $D(f)(x)=D(f(x))-f(D(x))$ we have then that $D(T)(x)=D(f)(x)q+f(x)D(q)$ as required.

Reverting to our previous notation to continue the discussion, we have shown that $D(\sum v_i\otimes f_i)=0$ as previously asserted.

For future reference, we assign distinct symbols for γ for each of the three identifications of the target (with a chosen projective coordinate system v_i, f_i, $1 \leq i \leq n$ on V):

$$\gamma_1 : W^* \to \mathrm{Hom}_R(V,W)^* \otimes_R V^* \text{ by } \phi \mapsto \sum m_{\phi,v_i} \otimes f_i$$

$$\gamma_2 : W^* \to \mathrm{Hom}_R(W,V) \otimes_R V^* \text{ by } \phi \mapsto \sum S_{\phi,v_i} \otimes f_i$$

$$\gamma_3 : W^* \to W^* \otimes_R V \otimes_R V^* \text{ by } \phi \mapsto \sum \phi \otimes v_i \otimes f_i$$

Under appropriate hypotheses, these maps are injective. It suffices to consider γ_3. Consider the map $E : V \otimes V^* \to R$ by $v \otimes f \mapsto f(v)$ which sends $\sum v_i \otimes f_i$ to $\sum f_i(v_i)$. As we noted above, $\sum v_i \otimes f_i$ is independent of the choice of a coordinate system. If V is free of rank r, we can choose e_i to be a basis and the h_i to be the dual basis, and hence $\sum f_i(v_i) = \sum h_i(e_i)$. Since V is locally free, we have this equality locally, and so if V is of constant rank r we have it globally. Thus, for $1 \otimes E : W^* \otimes V \otimes V^* \to W^*$, we have $(1 \otimes E)(\gamma_3(\phi)) = r\phi$ provided V is of constant rank r. So if r is invertible in R, $r^{-1}(1 \otimes E)$ is a left inverse for γ_3 and hence the latter is injective. Even if r is only a non-zero divisor on R, we can draw the same conclusion.

We will take this construction one further step. Let $S = S_R(\mathrm{Hom}_R(W,V))$ be the symmetric algebra, which we regard as a differential R algebra using the differential structure on $\mathrm{Hom}_R(W,V)$. The differential embedding $\mathrm{Hom}_R(W,V) \to S$ induces a differential embedding $\mathrm{Hom}_R(W,V) \otimes_R V^* \to S \otimes_R V^*$ and γ_2 induces a differential map $W^* \to S \otimes_R V^*$, which we call υ_2. Note that both $S \otimes_R W^*$ and $S \otimes_R V^*$ are differential S modules, and that $\Upsilon_2 : S \otimes_R W^* \to S \otimes_R V^*$ by $s \otimes \phi \mapsto s(\upsilon_2(\phi))$ is a homomorphism of differential S modules. Using γ_1 and

γ_3 we can similarly construct maps $\Upsilon_1 : S_R(\mathrm{Hom}_R(V,W)^*) \otimes_R W^* \to S_R(\mathrm{Hom}_R(V,W)^*) \otimes_R V^*$ and $\Upsilon_3 : S_R(W^* \otimes_R V) \otimes_R W^* \to S_R(W^* \otimes_R V) \otimes_R V^*$.

Although the Υ_i are differential homomorphisms of differential modules, they are also R module homomorphisms which are defined for any (finitely generated, projective) R modules. We are going to examine them in the special case that, as R modules, $V = W$ and that V is free with basis $v_1, \ldots, v_n$ and dual basis $f_1, \ldots, f_n$. This covers the case of differential modules whose underlying R modules are free of the same rank. We focus on the case $i = 2$. We use the notations already introduced.

Lemma 1.5 *Let R be a commutative ring and let V be a free R module with basis $v_1, \ldots, v_n$ and dual basis $f_1, \ldots, f_n$. Then*

$$\Upsilon_2 : S_R(Hom_R(V,V)) \otimes_R V^* \to S_R(Hom_R(V,V)) \otimes_R V^*$$
$$\text{by } s \otimes \phi \mapsto s(\upsilon_2(\phi))$$

is injective. Let $d = det(S_{f_i,v_j})$. Then $\Upsilon_2[d^{-1}]$ is an isomorphism.

Proof. Let $S = S_R(\mathrm{Hom}_R(V,V))$ and let $x_{ij} = S_{f_i,v_j}$. Then S is a polynomial ring over R in the n^2 elements x_{ij}. Any element of $S \otimes V^*$ can be written uniquely as $s = \sum_i s_i \otimes f_i$. The image of $s_i \otimes f_i$ is $s_i(\sum_j x_{ij} \otimes f_j)$. Thus the image of s is $\sum_i s_i(\sum_j x_{ij} \otimes f_j) = \sum_j(\sum_i s_i x_{ij}) \otimes f_j$; Let t_j denote $\sum_i s_i x_{ij}$ so that the image of s is $\sum_j t_j \otimes f_j$. Let $\underline{s} = (s_1, \ldots, s_n)$, $\underline{t} = (t_1, \ldots, t_n)$ (row vectors of length n), and let $X = [x_{ij}]$ ($n \times n$ matrix). Then $\underline{t} = \underline{s}X$. Since the entries of X are indeterminants over R, they are linearly independent over R. Thus, if $\underline{t} = 0$, $\underline{s} = 0$, which proves the assertion of injectivity. If we adjoin d^{-1} to S, then X is invertible in S, and hence we can conclude the isomorphism assertion.

We also have versions of Lemma 1.5 where Υ_2 is replaced by Υ_1 or Υ_3, and $S_R(\mathrm{Hom}_R(V,V))$ is replaced by $S_R(\mathrm{Hom}_R(V,V)^*)$ and $S_R\mathrm{Hom}_R(V^*,V))$, respectively. In the former case, x_{ij} is replaced by m_{f_i,v_j} and in the latter by $f_i \otimes v_j$.

In Lemma 1.5 and its variants we are only considering the projective module V. We have previously established that the morphisms Υ_i are differential, so as they stand they are assertions about the differential module V. Throughout this section, and elsewhere, we have been referring to the differential modules (W,D_W) and (V,D_V) by their first entries. We now want to apply Lemma 1.5 to differential modules whose underlying projective modules are the same: a pair (V,D_1) and (V,D_2). Then we have the differential map $\Upsilon_2 : S\otimes(V,D_2)^* \to S\otimes(V,D_1)^*$ where $S = S_R(\mathrm{Hom}_R((V,D_2),(V,D_1)))$. (Here (V,D_1) is playing the role of the differential module V and (V,D_2) is playing the role of the differential module W; we are using the ordered pair notation for the differentiations. The R modules remain the same.)

We are going to restate Lemma 1.5 as corollaries to itself where we use the above notation.

Corollary 1.6 *Let R be a differential commutative ring and let V be a free R module with basis $v_1,\ldots,v_n$ and dual basis $f_1,\ldots,f_n$. Let (V,D_1) and (V,D_2) be differential R modules. Let S be the differential R algebra $S_R(Hom_R((V,D_2),(V,D_1)))$ Then the differential map*

$$\Upsilon_2 : S\otimes_R(V,D_2)^* \to S\otimes_R(V,D_1)^* \text{ by } s\otimes\phi \mapsto s(\upsilon_2(\phi))$$

is injective. Let $d = det(S_{f_i,v_j})$. Then $\Upsilon_2[d^{-1}]$ is an isomorphism.

For later use, we also restate Lemma 1.5 for Υ_1.

Corollary 1.7 *Let R be a differential commutative ring and let V be a free R module with basis $v_1, \ldots, v_n$ and dual basis $f_1, \ldots, f_n$. Let (V, D_1) and $V, D_2)$ be differential R modules. Let S be the differential R algebra $S_R(Hom_R((V, D_1), (V, D_2))^*)$ Then the differential map*

$$\Upsilon_1 : S \otimes_R (V, D_2)^* \to S \otimes_R (V, D_1)^* \text{ by } s \otimes \phi \mapsto s(\upsilon_1(\phi))$$

is injective. Let $d = det(m_{f_i, v_j})$. Then $\Upsilon_1[d^{-1}]$ is an isomorphism.

The modules (V, D_i) and the various versions of the algebras S in Corollary 1.6 and Corollary 1.7, without their derivations, are all the same, namely R^n and $S_R(M_n(R))$. This means that for any pair of derivations D_1, D_2 on R^n, the induced derivation on $S = S_R(M_n(R))$ makes the two differential modules isomorphic after scalars are extended to $S[d^{-1}]$.

1.7 DIFFERENTIAL FREE MODULES

On the free R module R^n (which we think of as column n vectors), we have the coordinate–wise derivation $D_R((x_i)) = (D_R(x_i))$, which we will denote either D_0 or $(\cdot)'$. This makes R^n a differential R module. In particular, the R module $M_{m,n}(R)$ of $m \times n$ matrices over R is thus a differential module under this coordinate wise differentiation. If A is an $m \times n$ R matrix and B is an $n \times p$ R matrix, it is again a straightforward calculation that $(AB)' = A'B + AB'$. Moreover, with respect to transpose, we have $(A^T)' = (A')^T$. Since transpose is additive and R linear, the preceding equation implies that transpose is a differential isomorphism from $M_{m,n}(R)$ to $M_{n,m}(R)$.

Suppose (R^n, D_1) is a differential module. By Proposition 1.10, $D_1 - D_0 = T$ is an R endomorphism of R^n. Using the standard basis of R^n, there is an $n \times n$ matrix A with entries in R such that $Tx = Ax$ for all $x \in R^n$. Thus $D_1(x) = x' + Ax$ for all $x \in F^n$. Conversely, if we start with an $n \times n$ matrix A over R, then again by Proposition 1.10, the map $x \mapsto x' + Ax$ defines a differential structure on R^n. We are going to name this structure D_A. Note that in the case of the zero matrix 0, this notation coincides with our previous interpretation of D_0.

We use R^n for column n tuples, namely $M_{n,1}(R)$. We can also consider row n tuples, namely $M_{1,n}(R)$. (Sometimes $M_{1,n}(R)$ is denoted $R^{(n)}$.) By repeating the argument in the previous paragraph, we see that if $(M_{1,n}(R), D)$ is a differential module, then there is an $n \times n$ matrix A such that $D(x) = x' + xA$ for all $x \in M_{1,n}(R)$. We will call this differential structure ${}_AD$; as before, the preceding formula defines a differential structure for any given matrix. The transpose $M_{1,n}(R) \to M_{n,1}(R)$ is an R isomorphism which induces a differential isomorphism $(M_{1,n}(R), {}_AD) \to (M_{n,1}(R), D_{A^T})$.

We consider in more notational detail differential homomorphisms from R free differential modules to differential R modules. Because of the intended uses of this notation in Chapter 2, we consider the free module as row vectors. Thus, we suppose we have a differential R module W and an R linear map $f : (R^{(n)}, {}_AD) \to W$. As an R linear map, f is determined by where it sends the standard basis of $R^{(n)}$. So let $w_i \in W$ be the f image of e_i, where the latter is the row n tuple with i^{th} entry

1 and other entries 0. We write this in a "mixed" matrix notation as

$$f((x_1,\ldots,x_n)) = \sum_1^n x_i w_i = \begin{pmatrix} x_1, & x_2, & \ldots, & x_n \end{pmatrix} \begin{bmatrix} w_1 \\ w_2 \\ \ldots \\ w_n \end{bmatrix}.$$

For f to be differential we need that

$$f(D((x_1,\ldots,x_n)) = D(f((x_1,\ldots,x_n))).$$

To compute $D(f((x_1,\ldots,x_n)))$ we differentiate $\sum x_i w_i$:

$$D(\sum x_i w_i) = \sum D(x_i) w_i + \sum x_i D(w_i)$$

which in matrix form is

$$D(f((x_1,\ldots,x_n))) = D(\begin{pmatrix} x_1, & x_2, & \ldots, & x_n \end{pmatrix}) \begin{bmatrix} w_1 \\ w_2 \\ \ldots \\ w_n \end{bmatrix}$$

$$+ \begin{pmatrix} x_1, & x_2, & \ldots, & x_n \end{pmatrix} D(\begin{bmatrix} w_1 \\ w_2 \\ \ldots \\ w_n \end{bmatrix})$$

In this formalism $D((x_1,\ldots,x_n)) = (D(x_1),\ldots,D(x_n))$ and

$$D(\begin{bmatrix} w_1 \\ w_2 \\ \ldots \\ w_n \end{bmatrix}) = \begin{bmatrix} D(w_1) \\ D(w_2) \\ \ldots \\ D(w_n) \end{bmatrix}$$

so

$$D(f((x_1,\ldots,x_n))) = \begin{pmatrix} D(x_1), & D(x_2), & \ldots, & D(x_n) \end{pmatrix} \begin{bmatrix} w_1 \\ w_2 \\ \ldots \\ w_n \end{bmatrix}$$

$$+ \begin{pmatrix} x_1, & x_2, & \ldots, & x_n \end{pmatrix} \begin{bmatrix} D(w_1) \\ D(w_2) \\ \ldots \\ D(w_n) \end{bmatrix}$$

Now

$$D((x_1,\ldots,x_n)) = (D(x_1),\ldots,D(x_n)) + (x_1,\ldots,x_n)A$$

so applying f gives

$$f(D((x_1,\ldots,x_n))) =$$

$$(\begin{pmatrix} D(x_1), & D(x_2), & \ldots, & D(x_n) \end{pmatrix} + \begin{pmatrix} x_1, & x_2, & \ldots, & x_n \end{pmatrix} A) \begin{bmatrix} w_1 \\ w_2 \\ \ldots \\ w_n \end{bmatrix}.$$

Thus for f to be differential we need that

$$\begin{pmatrix} x_1, & x_2, & \ldots, & x_n \end{pmatrix} D(\begin{bmatrix} w_1 \\ w_2 \\ \ldots \\ w_n \end{bmatrix}) = \begin{pmatrix} x_1, & x_2, & \ldots, & x_n \end{pmatrix} A \begin{bmatrix} w_1 \\ w_2 \\ \ldots \\ w_n \end{bmatrix}.$$

for all tuples $(x_1, \ldots, x_n)$, hence that

$$D(\begin{bmatrix} w_1 \\ w_2 \\ \ldots \\ w_n \end{bmatrix}) = A \begin{bmatrix} w_1 \\ w_2 \\ \ldots \\ w_n \end{bmatrix}.$$

If $a_{i1}, \ldots, a_{in}$ are the entries of row i of A, this last equation is

$$D(w_i) = \sum_{j=1}^{n} a_{ij} w_j \text{ for all } j$$

So that we can refer to these calculations later, we restate them as a proposition.

Proposition 1.12 *Let $A = [a_{ij}]$ be an $n \times n$ matrix over R. Let W be a differential R module. Let $w_i \in W$ for $1 \le i \le n$. Then*

$$(R^{(n)}, {}_A D) \to W \text{ by } (x_1, \ldots, x_n) \mapsto \sum x_i w_i$$

is differential if and only if

$$D(w_i) = \sum_{j=1}^{n} a_{ij} w_j \text{ for all } j$$

Symbolically, if and only if

$$D(\begin{bmatrix} w_1 \\ w_2 \\ \ldots \\ w_n \end{bmatrix}) = \begin{bmatrix} D(w_1) \\ D(w_2) \\ \ldots \\ D(w_n) \end{bmatrix} = A \begin{bmatrix} w_1 \\ w_2 \\ \ldots \\ w_n \end{bmatrix}.$$

There is an isomorphism Φ of $M_{1,n}(R)$ to $\mathrm{Hom}_R(M_{n,1}(R), R)$ via $x \mapsto (y \mapsto xy)$. Let A be an $n \times n$ R matrix, put the D_A differential structure on $M_{n,1}(R)$, and the D_0 differential structure on R.

The resulting differential structure on $\mathrm{Hom}_R(M_{n,1}(R), R)$ is $D(f)(y) = D_0(f(y)) - f(D_A(y))$. If f is given by left multiplication by $x \in M_{1,n}(R)$, this formula becomes $(xy)' - x(y' + Ay) = x'y - xAy$. In other words, $D(\Phi(x)) = \Phi(x' - xA)$. It follows that Φ is a differential isomorphism of $(M_{1,n}(R), _{-A}D)$ to $\mathrm{Hom}_R(M_{n,1}(R), R)$, where the latter has the differential structure from D_A on $M_{n,1}(R)$ and D_0 on R. Thus the differential dual of $(M_{n,1}, D_A)$ is differentially isomorphic to $(M_{1,n}(R), _{-A}D)$. Also, we know that $(M_{1,n}(R), _{-A}D)$ is differentially isomorphic, via transpose, with $(M_{n,1}(R), D_{-A^T})$.

We can also consider the differential dual of (R^n, D_A) using a basis. Let $e_1, \ldots, e_n$ be the standard basis of R^n (e_i is the column n-tuple with i^{th} entry 1 and the rest 0) and let $\hat{e}_i$, $1 \leq i \leq n$ be the corresponding dual basis of $\mathrm{Hom}_R(R^n, R)$ ($\hat{e}_i(e_j) = \delta_{i,j}$). Then $D(\hat{e}_i)(e_j) = D_R(\hat{e}_i(e_j)) - \hat{e}_i(D_0(e_j) + Ae_j)$. Since $D_R(\hat{e}_i(e_j))$ and $D_0(e_j)$ are the zero scalar and vector, respectively, $D(\hat{e}_i)(e_j) = -\hat{e}_i(Ae_j)$. Thus $D(\hat{e}_i)(e_j)$ is the negative of the entry a_{ij} in row i and column j of A. In terms of the basis $\hat{e}_i$, there are elements $y_1, \ldots, y_n$ of R such that $D(\hat{e}_i) = y_1\hat{e_1} + \cdots + y_n\hat{e_n}$. Evaluating both sides of this equation at e_j shows that $y_j = -a_{ij}$. The basis $\hat{e}_i$ identifies $\mathrm{Hom}_R(R^n, R)$ with R^n, with $\sum x_i\hat{e}_i$ corresponding to the column vector with entries y_i. Then this last equation identifies D with D_{-A^T}. Thus the differential dual of (R^n, D_A) is again $(R^n, -D_{A^T})$ (unnaturally, since it depends on the choice of basis, even though the choice is the standard one).

We record the various isomorphisms just observed in the following proposition:

Proposition 1.13 *Let $A \in M_n(R)$. Then $(R^{(n)}, {}_A D)$ is differentially isomorphic to (R^n, D_{A^T}) by the transpose, and the usual matrix multiplication isomorphism*

$$R^n \to Hom_R(R^n, R) \text{ by } \underline{x} \mapsto (\underline{y} \mapsto \underline{xy}$$

is a differential isomorphism

$$(R^{(n)}, {}_{-A} D) \to Hom_R((R^n, D_A), R).$$

Suppose B is also an $n \times n$ matrix over R. We ask when R^n with differential structure D_B is isomorphic as a differential R module to R^n with differential structure D_A. For that to happen, we need an R isomorphic endomorphism T of R^n with $D_B(T(x)) = T(D_A(x))$ for all $x \in R^n$. If T is given by multiplication by the invertible $n \times n$ matrix S over R, this equation becomes $(Sx)' + BSx = Sx' + SAx$. Since $(Sx)' = S'x + Sx'$, this equation becomes $S'x + BSx = SAx$, or $S'x = SAx - BSx$. Since this holds for all x, we conclude that

Proposition 1.14 *The $n \times n$ matrices A and B over R give isomorphic differential structures on R^n if and only if there is an invertible $n \times n$ matrix S over R such that $S' = SA - BS$. In particular, R^n with the differential structure D_B is isomorphic to R^n with coordinate-wise derivation (case where $A = 0$) if and only if there is an invertible S with $S' = -BS$ and $R^{(n)}$ with the differential structure ${}_B D$ is isomorphic to $R^{(n)}$ with coordinate-wise derivation if and only if there is an invertible S with $S' = -SB$.*

The final clause in Proposition 1.14 follows from the penulimate one by taking transposes.

Of course, R^n with the differential structure D_B is isomorphic to R^n with the differential structure D_0 if and only if it has a basis $x_1,\dots,x_n$ with $D_B(x_i)=0$, $1\le i\le n$. Let S be the $n\times n$ matrix whose column i is x_i. Note that S is invertible since the x_i are a basis. Since $D_B(x_i)=x_i'+Bx_i$, $S'+BS=0$, or $S'=-BS$. This is an equivalent version of the first clause of the "in particular" assertion of Proposition 1.14.

Suppose that (R^n,D_B) does have a basis of constants $x_1,\dots,x_n$ and that $X\in GL_n(R)$ is the matrix whose columns are $x_1,\dots,x_n$. Let $Y=X^{-1}$ and let $y_1,\dots,y_n\in R^{(n)}$ be the rows of Y. Note that $y_ix_j=\delta_{ij}$, so that under the isomorphism $(R^{(n)},_{-B}D)\to \mathrm{Hom}_R((R^n,D_B),R)$ of Proposition 1.13 the y_i's map to the dual basis of the x_j's. We differentiate the equation $YX=I_n$ and find $Y'X+Y(-BX)=0$, or $Y'X=YBX$, so $Y'=YB$. This tells us that $y_1,\dots,y_n$ is a basis of constants for $(R^{(n)},_{-B}D)$.

The minus sign in the equation $S'=-BS$ presents no mathematical difficulty; it is simply a consequence of using D_A to stand for the derivation $x\mapsto x'+Ax$ on R^n. Besides any a esthetic considerations, when we consider the equivalent version of Proposition 1.14 just mentioned, the criteria for x to be a constant is $x'=-Ax$, which has the same minus sign. Because of considerations like these, there is an alternative notation: for an $n\times n$ matrix A,

$$(R^n,\partial_A) \text{ denotes the differential module } \partial_A(x)=x'-Ax.$$

Note that $(R^n,\partial_A)=(R^n,-D_A)$. Later, we will have need of the notation $(R^{(n)},{}_A\partial):=(R^{(n)},-{}_AD)$.

Then $y\in R^n$ is a constant for ∂_A provided $y'=Ay$ and $x\in R^{(n)}$ is a constant for ${}_A\partial$ provided $x'=xA$. The final sentence of Proposition

1.14, and the discussion of the dual after the statement, then takes the form:

Corollary 1.8 *R^n with the differential structure ∂_A is isomorphic to R^n with coordinate-wise derivation (i.e., (R^n, ∂_A) has a basis of constants) if and only if there is an invertible $n \times n$ matrix T with $T' = AT$, and $R^{(n)}$ with the differential structure ${}_A\partial$ is isomorphic to $R^{(n)}$ with coordinate-wise derivation (i.e., $(R^{(n)}, {}_A\partial)$ has a basis of constants) if and only if there is an invertible $n \times n$ matrix T with $T' = TA$. If there is an invertible matrix X with $X' = AX$, then the columns of X are a basis of constants for (R^n, ∂_A) and the rows of X^{-1} are a basis of constants for $(R^{(n)}, {}_A\partial)$.*

(We switched the notation to T from S and introduced X to avoid confusion in the application of Corollary 1.6.)

Given A in $M_n(R)$, there may or may not be an invertible matrix $T \in M_n(R)$ satisfying $T' = AT$. But we can produce one in a differential extension ring of R. Referring to the notation of Corollary 1.1, on the polynomial $F[x_{11}, \ldots, x_{nn}]$ define the derivation such that $X' = AX$ (where $X = [x_{ij}]$). Let T denote X in the ring $S = F[x_{11}, \ldots, x_{nn}][\det(X)^{-1}$. Then T is invertible and $T' = AT$. We apply Corollary 1.8.

Corollary 1.9 *Let $A \in M_n(R)$. Let $S = F[x_{11}, \ldots, x_{nn}][det(X)^{-1}]$ with derivation $X' = AX$. Then $(S^n, \partial_A) = S \otimes_R (R^n, \partial_A)$ has a basis of constants formed by the columns of X.*

Corollary 1.9 is, in fact, a restatement of Corollary 1.6: the rings S in both are isomorphic. But in some sense Corollary 1.6 is the more natural, since it ultimately derives from the evaluation homomorphism

$\mathrm{Hom}_R(V,W)\otimes V \to W$, which in the case here, where both differential modules are R^n as R modules is the basic multiplication map $M_n(R)\otimes R^n \to R^n$ by $A\otimes x \mapsto Ax$.

Proposition 1.14 and Corollaries 1.8 and 1.9 are the result of considering isomorphisms of the differential module (R^n, D_A) (or (R^n, ∂_A)) with (R^n, D_0). The conclusions, however, are just statements about matrices. So we are going to look at those strictly from the matrix point of view. In that context, we are looking at an invertible matrix S (Proposition 1.14) or T (Corollary 1.8) satisfying $S' = -BS$ or $T' = AT$. Starting with the modules, that is, starting with B or A, we are looking for S or T. Corollary 1.8 tells how to find the invertible matrix. On the other hand, if we start with the invertible matrix, then $-B$ or A are $S'S^{-1}$ or $T'T^{-1}$. There are some interesting consequences of considering such forms.

Thus, suppose $X \in M_n(R)$ is invertible. Differentiating the equation $X^{-1}X = I_n$ gives $(X^{-1})'X + X^{-1}X' = 0$ or $(X^{-1})'X = -X^{-1}X'$. Multiplying on the right by X^{-1} then gives

$$(X^{-1})' = -X^{-1}X'X^{-1}.$$

Let $A = X'X^{-1}$, so $X' = AX$. Suppose Z is an $n\times p$ matrix with $Z' = AZ$. Consider $C = X^{-1}Z$. Its derivative is $-X^{-1}X'X^{-1}Z + X^{-1}Z'$ or $-X^{-1}AZ + X^{-1}AZ = 0$. Thus $C' = 0$, which means $C \in M_{n,p}(R^D)$. Since $C = X^{-1}Z$, this means $Z = XC$, where C is a matrix of constants. Conversely, if $Z = XC$ where $C' = 0$, then $Z' = (XC)' = X'C + XC' = AZ$. We summarize:

Proposition 1.15 *Let $X \in M_n(R)$ be invertible. Let $A = X'X^{-1}$. Then $X' = AX$, and*

$$\{Z \in M_{n,p}(R) | Z' = AZ\} = XM_{n,p}(R^D).$$

When R is a field, so is R^D and then Proposition 1.15 with $p = 1$ says the following:

Corollary 1.10 *Let F be a differential field. Suppose that $A \in M_n(F)$ is such that there exists $X \in GL_n(F)$ with $X' = AX$. Then the columns of X form a basis over F^D for $\{x \in F^n | x' = Ax\}$.*

Proof. Proposition 1.15 asserts that if $x' = Ax$, then x is an F^D linear combination of columns of X. Since X is invertible, these columns are linearly independent in F^n. Since the unit vectors e_i with entry 1 in row i and others 0 are in $M_{n,1}(F^D)$ so Proposition 1.15 also asserts that Xe_i, column i of X, is in the space in question.

1.8 CYCLIC MODULES

A differential R module M is *cyclic* if there is an element $a \in M$ such that $M = \sum RD^i(a)$; a is called a cyclic generator of M. For any differential module M and any $a \in M$, the differential submodule $M = \sum RD^i(a)$ is a cyclic differential module. We will be most interested in the case where M is finitely generated as an R module, in which case if M is cyclic $M = \sum_{i=0}^{n} RD^i(a)$ for some n. We will show in Chapter 2 that when R is a field of characteristic zero whose subfield of constants is algebraically closed *proper* that all finitely generated differential modules are cyclic. For differential fields satisfying these criteria, then, differential submodules of cyclic modules, being differential modules, are also cyclic. As the following proposition shows, submodules of finite-dimensional cyclic modules are cyclic over any field.

We first make a remark about the form of a basis of a finite-dimensional cyclic module which will be used in the proof of Proposition 1.16 and elsewhere.

Lemma 1.6 *Let M be a cyclic differential module over the differential field F. Assume $dim_F(M) = n > 0$. If a is any generator of M then $D^i(a)$, $0 \leq i \leq n-1$ is a basis of M.*

Proof. Let k be maximal with respect to the property that $D^i(a)$, $0 \leq i \leq k$ are linearly independent. Note that since $M \neq \{0\}$, $a \neq 0$ so that k is at least 0. Then $D^{k+1}(a) \in \sum_{i \leq k} RD^i(a)$, say $D^{k+1}(a) = \sum_{i \leq k} a_i D^i(a)$. Since $D(a_i D^i(a) = D(a_i)D^i(a) + a_i D^{i+1}(a)$, we see that $D^{k+2}(a) = D(D^{k+1}(a)) \in a_k D^k(a) + \sum_{i \leq k} RD^i(a)$, and since $D^k(a) \in \sum_{i \leq k} RD^i(a)$, this shows that $D^{k+2}(a) \in \sum_{i \leq k} RD^i(a)$. Continuing inductively, we see that $D^j(a) \in \sum_{i \leq k} RD^i(a)$ foe all j, so that $M = \sum_{i \leq k} RD^i(a)$. Thus the linearly independent $D^i(a)$, $0 \leq i \leq k$ span M, so $k = n-1$ and $D^i(a)$, $0 \leq i \leq n-1$ is a basis.

Proposition 1.16 *Let F be a differential field, let M be a cyclic differential F module of finite dimension as an F vector space, and let $N \subset M$ be a differential submodule. Then N is a cyclic differential module.*

Proof. By Lemma 1.6 M has a basis of the form $D^i(a)$, $0 \leq i \leq n-1$. Any element $x \in M$ can be expressed uniquely as $x = x_{n-1}D^{n-1}(a) + \cdots + x_1 D(a) + x_0 a$. If $x \neq 0$, define the *degree* of x to be the maximum

i such that $x_i \neq 0$ and let $\deg(x)$ denote the degree, if $x = 0$ define $\deg(x) = -\infty$. If $N = \{0\}$, then N is cyclic with cyclic generator 0. Otherwise, let $b \in N$ with $b = \sum b_i D^i(a)$ be an element of least non-negative degree k. We multiply b by the inverse of b_k, so we can assume $b_k = 1$. Suppose $c \in N$ also has degree k. Then $c - c_k b$ has degree smaller than k, so $c - c_k b = 0$, so $c = c_k b$, and hence c belongs to the cyclic submodule $N_0 = \sum RD^i(b)$ generated by b. Suppose we have shown that every element N of degree $k, k+1, \ldots, m-1$ belongs to N_0 and that $x = x_m D^m(a) + \cdots + x_0$ is an element of N of degree m. Let $p = m - k$. For $i < k$, $\deg(D^p(b_i D^i(a))) < m$. Thus $\deg(x - x_m D^p(b)) < m$, so by assumption $x - x_m D^p(b)$ belongs to N_0 and hence so does x. It then follows by induction that $N = N_0$ is cyclic and generated by b.

CHAPTER 2

Differential Modules over Differential Fields

2.1 DIFFERENTIAL GALOIS THEORY

We will carry out the study of differential modules over differential fields using the tools of differential Galois theory. From some points of view, differential Galois theory *is* the theory of differential modules over a differential field. In those perspectives, one starts with the tensored category of finite-dimensional differential modules over a differential field (hypotheses are necessary on the subfield of constants) and applies the theory of Tannakian categories to produce the differential Galois group and to prove the fundamental Galois correspondence theorem.

In the perspective adopted here, Differential Galois Theory is about finding solutions of (monic, linear homogeneous, ordinary) differential equations in differential (extension) fields, in the same sense that ordinary (henceforth polynomial) Galois theory is about roots of a polynomial in extension fields. In fact, however, as Don Knutson once quipped,

DOI: 10.1201/9781003452164-2

[polynomial] Galois Theory is about *not* solving [polynomial] equations. As far as applications go, this may appear to be true of differential Galois theory as well: probably the best known and most often utilized of its applications is to establish the non-integrability of certain physical systems. And it is true that Galois's initial big achievement for his theory was the demonstration that the roots of polynomial of a degree five or larger could not be expressed as radicals of rational algebraic expressions in the coefficients of the polynomial. Modern (mid twentieth century to the present) polynomial Galois Theory is usually presented as a theory of (algebraic, normal, separable) field extensions, with the fundamental theorem being that the lattice of subfields of the extension is inversely isomorphic to the lattice of closed subgroups of the group of all field automorphisms of the extension field which are the identity on the base, and where the the normal subgroups correspond to the normal subextensions. Polynomials and their roots have disappeared from this formulation, although not necessarily from the proofs. This sort of conceptual Galois Theory has been conceived and applied in many other branches of algebra. It remains useful, nonetheless, given a field and a (monic) polynomial equation over it, to consider what the algebraic rational relations the theoretical roots of the equation might have, what we might call traditional polynomial Galois theory, and it is the analogous formulation that we here call differential Galois theory. To be specific, let k be the base field for a polynomial Galois extension (which we will take to be characteristic 0, so no issues of inseparability arise), let $f \in k[x]$ be a separable monic polynomial of degree $n > 1$, and let $E \supseteq k$ be a splitting field for f over k. Let $A = \{\alpha_1, \ldots, \alpha_n\}$ be the set of roots of f in E; then $E = k(\alpha_1, \ldots, \alpha_n)$. In $E[x]$ $f = \prod_i (x - \alpha_i)$. The group $G = \mathrm{Aut}_k(E)$

is represented faithfully as permutations of A. Then $E \supseteq k$ is a Galois extension with group G, and all finite normal separable extensions (that is, Galois extensions) of k are of this form for suitable f.

The results of differential Galois theory to be summarized here mostly come from **Lectures on Differential Galois Theory**, University Lecture Series 7, American Mathematical Society, Providence 1994 (henceforth ULECT7).

There are many presentations of the differential Galois theory: Kolchin, E. R. **Differential algebra and algebraic groups**,Pure Appl. Math., Vol. 54 Academic Press, New York-London, 1973, the definitive account; Kaplansky, Irving **An introduction to differential algebra** Second edition Publications de l'Institut de Mathématique de l'Université de Nancago, No. V Actualités Sci. Indust., No. 1251 [Current Scientific and Industrial Topics] Hermann, Paris, 1976, a classic brief introduction; van der Put, Marius and Singer, Michael F. **Galois theory of linear differential equations** Grundlehren Math. Wiss., 328, Springer-Verlag, Berlin, 2003., a comprehensive modern account; and Crespo, Teresa; Hajto, Zbigniew **Algebraic groups and differential Galois theory** Grad. Stud. Math., 122 American Mathematical Society, Providence, RI, 2011, probably the best modern textbook. Pierre Deligne showed how differential Galois theory is a special case of Tannakian Category theory in Deligne, P., **Catégories tannakiennes**. The Grothendieck Festschrift, Vol. II, 111–195, Progr. Math., 87 Birkhäuser Boston, Inc., Boston, MA, 1990; a brief accessible account of this theory can be found in Daniel Bertrand's review of ULECT7 in the Bulletin (New Series) of the American Mathematical Society, Volume 33, Number 2, April 1996.

We fix here and throughout this chapter a differential field F with derivation $D = D_F$ whose subfield $C = F^D$ of constants is assumed to be algebraically closed and of characteristic zero. A (monic, linear, homogeneous) *differential operator* of order $n \geq 1$ over F is an expression of the form $L(Y) = Y^{(n)} + a_{n-1}Y^{(n-1)} + \cdots + a_1Y + a_0Y^{(0)}$. Let $E \supseteq F$ be a differential extension field and let $y \in Y$. Then $L(y) \in E$ is defined to be $D_E^n(y) + a_{n-1}D_E^{n-1}(y) + \cdots + a_1D_E(y) + a_0y$. (When there is no confusion, we will denote D_E by D.) We call the expression $L(Y) = 0$ the differential equation associated with $L(Y)$ and the set $\{y \in E | L(y) = 0\}$ the set of solutions in E of the differential equation. We also use L to denote the operator $L(Y)$ so that the differential equation is $L = 0$, and we also write the set of solutions as $L^{-1}(0)$. The latter is a vector space over the constants of E; it is a theorem that the dimension of this vector space is at most n, the order of L. In direct analogy with the splitting field of a polynomial, we define a *Picard–Vessiot* extension of F for L to be a differential field extension of F such that

1. The subfield of constants of E is C.

2. Let $V = L^{-1}(0)$ in E. Then $\dim_C(V) = n$.

3. E is generated over F as a differential field by V.

We say that if $E \supseteq F$ satisfies the first condition, then it is a *no new constants extension*, and that if it satisfies the second condition, then it contains a *full set of solutions* of L. Thus a Picard–Vessiot extension of F for L is a no-new-constants extension generated by a full set of solutions for L. It is a theorem that Picard–Vessiot extensions exist for every F

and every L over F, and that any two are isomorphic as differential field extensions of F. Briefly, Picard–Vessiot extensions exist and are unique.

Given L, a Picard–Vessiot extension of F for L is constructed as follows: on the polynomial ring $S = S_L = F[X_{ij} | 0 \leq i \leq n-1 \, , 1 \leq j \leq n]$ define a derivation D extending the derivation D_F by $D(X_{ij}) = X_{i+1,j}$ if $i < n-1$ and $D(X_{n-1,j}) = -a_{n-1}X_{n-1,j} - \cdots - a_1 X_{1,j} - a_0 X_{0,j}$. Notice that this makes $X_{0,j}$ a solution of $L(Y) = 0$ in the polynomial ring T and hence in its quotient field K. Moreover, $\{X_{01}, \ldots, X_{0n}\}$ differentially generate S as a differential F algebra and hence K as a differential field extension of F. And these elements are even algebraically independent over F and hence certainly linearly independent over C. In general, however, $K \supset F$ will not be a no new constants extension. To remedy this, we apply a theorem which asserts that the quotient field of any affine differential algebra over F which has no proper differential ideals (such an algebra is always an integral domain) is a no-new-constants extension of F. We can create such a ring by taking the quotient of S by a maximal differential ideal. The quotient field of this ring then is a differential field extension of F which is a no new constants extension generated by the n solutions of L which are the images of the $X_{0,j}$. However, these images may no longer be linearly independent over C. We can remedy this by going back to S and considering the polynomial $d = \det(X_{ij})$ which is the Wronskian determinant of $\{X_{01}, \ldots, X_{0n}\}$. (It will be convenient later to have a notation for Wronskians: if $z_1, \ldots, z_m$ are elements of a differential ring, $w(z_1, \ldots, z_m)$ denotes the determinant of the $m \times m$ matrix $W(z_1, \ldots, z_n)$ whose i^{th} row is D^{i-1} applied to $z_1, \ldots, z_n$.)

Let $R = R_L = S[d^{-1}]/P$ where P is a maximal differential ideal of $S[d^{-1}]$. This latter is an affine F algebra, so the quotient $E = E_L$ of R

is a no new constants extension of F, and the images x_{0j} of the X_{0j} generate E differentially over F. The image of d in E, the Wronskian of $\{x_{01},\ldots,x_{0n}\}$, is a unit (that is, non-zero) so $\{x_{01},\ldots,x_{0n}\}$ is a full set of solutions of $L=0$ in E. This means that $E \supseteq F$ is a Picard–Vessiot extension of F for L. Because of the uniqueness of Picard–Vessiot extensions, every Picard–Vessiot extension for every L and every F can be regarded as constructed this way.

When there are several operators under discussion, we use the notations S_L, R_L, and E_L; if the ground field F needs to be explicitly identified, we write $S_{L,F}$, $R_{L,F}$, and $E_{L,F}$; when the operator is clear from context, we just write S, R, and L, as we have been doing heretofore.

Differential field isomorphisms extend to Picard–Vessiott extensions. Since this fact is not proved explicitly in ULECT7 in the form we intend to use it below, we provide a proof now.

Proposition 2.1 *Let $\mathcal{K}$ be a differential field with an algebraically closed characteristic zero subfield of constants. Let $L = D^n + a_{n-1}D^{n-1} + \cdots + a_1D + a_0$ be a differential operator over $\mathcal{K}$. Let $\sigma : \mathcal{K} \to K$ be a differential field isomorphism. Let $L^\sigma = D^n + \sigma(a_{n-1})D^{n-1} + \cdots + \sigma(a_1)D + \sigma(a_0)$ be the differential operator over K obtained from L by applying σ to its coefficients. Let $\mathcal{E}$ be a Picard–Vessiot extension of $\mathcal{K}$ for L, and let E be a Picard–Vessiot extension of K for L^σ. Then there is a differential field isomorphism $\tau : \mathcal{E} \to E$ such that $\tau|\mathcal{K} = \sigma$.*

Proof. By the uniqueness of Picard–Vessiot extensions, we have differential field isomorphisms $\mathcal{E} \to E_{L,\mathcal{K}}$ and $E_{L^\sigma,K} \to E$ which are the identity on $\mathcal{K}$ and K respectively. Thus is suffices to prove the proposition when $\mathcal{E} = E_{L,\mathcal{K}}$ and $E_{L^\sigma,K} = E$. Write $S_{L,\mathcal{K}} = \mathcal{K}[\chi_{01},\ldots,\chi_{n-1,n}]$ and write $S_{L^\sigma,K} = K[X_{01},\ldots,X_{n-1,n}]$ with derivations $D(\chi_{ij}) = \chi_{i+1,j}$ and $D(X_{i,j}) = X_{i+1,j}$ for $i < n-1$ and for $i = n-1$ $D(\chi_{n-1,j}) = -\sum_0^{n-1} a_i\chi_{i,j}$ and $D(X_{n-1,j}) = -\sum_0^{n-1} \sigma(a_i)X_{ij}$ on the variables extending the derivations on $\mathcal{K}$ and K. The ring differential ring isomorphism $S_{L,\mathcal{K}} \to S_{L^\sigma,K}$ passes to a differential ring isomorphism $\mu : S_{L,\mathcal{K}}[\delta^{-1}] \to S_{L^\sigma,K}[d^{-1}]$. Let Q be a maximal differential ideal of $S_{L,\mathcal{K}}[\delta^{-1}]$. Then $\mu(Q)$ is a maximal differential ideal of $S_{L^\sigma,K}$[-1] and μ passes to a differential ring isomorphism $S_{L,\mathcal{K}}[\delta^{-1}]/Q \to S_{L^\sigma,K}[d^{-1}]/\mu(Q)$, and then to a differential isomorphism of their quotient fields. This is the desired isomorphism τ.

Suppose we are only interested in a single solution of $L = 0$. Then we can consider the n variable polynomial ring $F[X_0,\ldots,X_{n-1}]$ with derivation $D(X_i) = X_{i+1}$, $0 \le i \le n-2$, and $D(x_{n-1}) = -\sum_{i=0}^{n-1} a_iX_i$. By construction $L(X_0) = 0$, so this will be true of the image of X_0 under any differential homomorphism from $F[X_0,\ldots,X_{n-1}]$ to a differential field $K_1 \supseteq F$. Suppose that K_1 is a no new constants extension of F, and let x_0 be the image of X_0 in K_1. Suppose that $K \subseteq K_1$ is a subfield containing F and that K_1 is generated as a differential field by K and x_0. Suppose that $K_2 \supseteq K_1$ is a Picard–Vessiot extension for L, here regarded as a differential operator with coefficients in K. In particular $K_2 \supseteq K$ is a no new constants extension of K, which is by assumption a no new constants extension of F, and hence $K_2 \supseteq F$ is a no new constants extension. Let $V = \{y \in K_2 | L(y) = 0\}$ be the set of solutions of $L = 0$ in K_2. Since $K_2 \supset K_1$ is Picard–Vessiot for L, $\dim_C(V) = n$. Since $L(x_0) = 0$, $x_0 \in V$. Consider the differential subfield of K_2 generated by K and V. This is a no new constants extension of K differentially generated by a full set of solutions of $L=0$, so it is a Picard–Vessiot extension of K for L, and it contains x_0 so it contains K_1.

We will need this construction later so we record it as a Lemma:

Lemma 2.1 *Let $L = D^n + a_{n-1}D^{n-1} + \cdots + a_1 D + a_0$ be a differential operator over F. Give the n variable polynomial ring $F[X_0, \ldots, X_{n-1}]$ the derivation D extending D_F with $D(X_i) = X_{i+1}$, $0 \leq i \leq n-2$, and $D(x_{n-1}) = -\sum_{i=0}^{n-1} a_i X_i$. Let $\mathcal{K}_1 \supseteq F$ be a no new constants extension and $\phi : F[X_0, \ldots, X_{n-1}] \to \mathcal{K}_1$ a differential F algebra homomorphism. Suppose $\mathcal{K}$ is a subfield of $\mathcal{K}_1$ which contains F and such that $\mathcal{K}_1$ is generated as a differential field over $\mathcal{K}$ by $a = \phi(x_0)$. Then there is a Picard–Vessiot extension $\mathcal{E} \supset \mathcal{K}$ with $\mathcal{K}_1 \subset \mathcal{E}$.*

We will consider the ring $R = R_L$ used to construct $E = E_L$ as a subring of E which contains F. It turns out to coincide with the set of all elements of E which satisfy a (monic, linear, homogeneous) differential equation over F.

Let $E \supseteq F$ be a Picard–Vessiot extension for L, and let $G = \mathrm{Aut}_F^{\mathrm{diff}}(E)$ be the group of differential field automorphisms of E which are the identity on F. Since G fixes the coefficients of L it preserves $V = L^{-1}(0)$ and acts C linear automorphism of V. Since V generates E as a differential field, the map $G \to GL(V)$ is injective. It is a theorem that the image of G is an algebraic subgroup of $GL(V) = GL_n(C)$. The fundamental theorem of differential Galois theory asserts that the differential fields K intermediate between F and E are precisely the fixed fields of (Zariski) closed subgroups of G (to K corresponds the group $G = \mathrm{Aut}_K^{\mathrm{diff}}(E)$. Further,

K itself is a Picard–Vessiot extension of F if and only if $H = \mathrm{Aut}_K^{\mathrm{diff}}(E)$ is a normal subgroup of G, and if it is G/H is isomorphic to $\mathrm{Aut}_F^{\mathrm{diff}}(K)$.

By construction, G acts rationally, indeed linearly, on V. Since $\{x_{01}, \ldots, x_{0n}\}$ is a basis of V, elements of G transform the matrix $[x_{ij}]$ by multiplying by a matrix in $GL_n(C)$ and hence transform d (the determinant of $[x_{ij}]$ by multiplication by scalars (that is, by a character of G). Thus G preserves R and acts rationally on it. In fact R is the set of all elements of E on which G acts rationally. Recall that R is also the subset of E of elements that satisfy differential equations. In fact we also have that for $y \in E$, $\dim_F(\sum_i FD^i y) < \infty$ if and only if $\dim_C(\sum_{\sigma \in G} C\sigma(y)) < \infty$. (The first finiteness condition is equivalent to being the solution of a differential equation, and the second to being acted on rationally by G.) Note that $\sum_i FD^i y$ is the F differential module generated by y and $\sum_{\sigma \in G} C\sigma(y)$ is the (C) G module generated by y. Both are contained in R. Suppose $y_1 \in R$ and let $y_1, \ldots, y_n$ be a C basis of the G module generated by y_1. One can show that the differential operator $L = w(y_1, \ldots, y_m)^{-1} w(Y, y_1, \ldots, y_n)$, which is of order m, monic, linear, and homogeneous, has coefficients in F, as they turn out to be fixed by G. Thus $y_1, \ldots, y_m$ are a full set of solutions of $L = 0$. The D modules $\sum_i D^i y_j$ are all isomorphic. Their sum is a differential module and a G module, and so is the differential G modules generated by y_1.

It is important to remember that G module consists of differential automorphisms so that the actions of G and D commute on E and R.

Since finite algebraic extensions of differential fields are (in characteristic zero) differential extensions, and in a unique way, the algebraic closure $\overline{F}$ of F is also a differential field. It is a theorem of Kolchin, actually a special case of a theorem of Kolchin, that the differential ring

$\overline{F} \otimes_F R$ as an $\overline{F}$ algebra and a ring with rational G action is isomorphic to $\overline{F} \otimes_C C[G]$ ($C[G]$ is the coordinate ring of G). Usually we write $\overline{F} \otimes_C C[G]$ as $\overline{F}[G]$. Similarly, $F[G]$ denotes $F \otimes_C C[G]$. If R has an F point (an F algebra homomorphism $R \to F$) then Kolchin's theorem becomes an isomorphism $R \to F[G]$.

Even without an F point, Kolchin's theorem implies that R is isomorphic to $F[G]$ as a rational G module: since both R and $F[G]$ are affine F algebras, the $\overline{F}$ algebra and G module isomorphism $\overline{F} \otimes_F R \to \overline{F}[G]$ can be realized as $h : K \otimes_F R \to K \otimes_F F[G]$ for some finite field extension $K \supset F$, say $[K : F] = n$. Since G acts trivially on K, h is a G module isomorphism $F^{(n)} \otimes_F R$ to $F^{(n)} \otimes_F F[G]$ from which it can then be deduced that R is isomorphic to $F[G]$. See Chinburg, Ted); Juan, Lourdes; Magid, Andy R. Picard-Vessiot extensions with specified Galois group, Pacific J. Math 243 (2009), no. 2, 233–242. This is the analogue, in differential Galois theory, of the Normal Basis Theorem in polynomial Galois theory.

Just as polynomial Galois theory has a version for infinite extensions, where the fields in question are unions of finite, normal, separable subextensions, the Galois groups are profinite, and the Fundamental Theorem gives a correspondence between intermediate fields and closed subgroups, there is a differential Galois theory for infinite extensions, where the fields in question are unions of Picard–Vessiot subextensions, the Galois groups are proalgebraic, and the Fundamental Theorem gives a correspondence between intermediate differential fields fields and closed subgroups. An especially important case is where every Picard–Vessiot extension of the base field has a copy in the extension. In this case we call the extension a Picard–Vessiot *closure*, although it is not Picard–Vessiot closed.

Most of the infinite-differential Galois theory is a straightforward extension from the Picard–Vessiot case. As we summarize the needed facts, we will point out where difficulties can arise.

We adopt the following notations: F^{PV} denotes a Picard–Vessiot closure of F (notation introduced by Anand Pillay). Thus

$$F^{PV} = \cup\{E \subset F^{PV} | E \supset F \text{ is a Picard–Vessiot extension}\}$$

and for every linear operator L with coefficients in F, there is a Picard–Vessiot extension of F for L contained in F^{PV}. If $E \subset F^{PV}$ is a Picard–Vessiot extension, let $G(F^{PV}/E)$ be the group of differential automorphisms of F^{PV} fixing E, and let $G(E/F)$ be the group of differential automorphisms of E fixing F; $\Pi_F = G(F^{PV}/F)$ maps to $G(E/F)$ by restriction with kernel $G(F^{PV}/E)$. The restriction maps are compatible and define a map $G(F^{PV}/F) \to \varprojlim_E G(E/F)$, where the limit is over $\{E \subset F^{PV} | E \supset F \text{ is a Picard–Vessiot extension}\}$, which is an isomorphism. Each $G(E/F)$ is an algebraic group over C and the restriction maps between them are morphisms of algebraic groups. Thus their inverse limit is a proalgebraic group. The subgroup $G(F^{PV}/E)$ of $G(F^{PV}/F)$ is closed and normal, with $(F^{PV})^{G(F^{PV}/E)} = E$. When we do need to specify the base field, we will write Π for Π_F.

For $E \supset F$ Picard–Vessiot and contained in F^{PV} we let $T(E/F)$ denote its Picard–Vessiot ring. Recall that $T(E/F)$ consists of all elements of E which are solutions of differential equations over F and well as all elements of E on which $G(E/F)$ acts rationally. Let $R = R^{PV} = T(F^{PV}/F)$ be the set of all elements of F^{PV} which satisfy differential equations over F. Then $R = \cup\{T(E/F)|E \supset$

F is a Picard–Vessiot extension}, which also shows that R is the set of all elements of F^{PV} on which $G(F^{PV}/F)$ acts rationally.

We may ask whether there is a Kochin's Theorem for R: that is, is there an $\overline{F}$ algebra and $G(F^{PV}/F)$ module isomorphism $\overline{F} \otimes_F R \to \overline{F}[G(F^{PV}/F)]$? One can be constructed from the direct limit of the isomorphisms $\overline{F} \otimes_F T(E/F) \to \overline{F}[E/F)]$, provided there are compatible $\overline{F}$ points $T(E/F) \to \overline{F}$ for all Picard–Vessiot extensions $E \supset F$ contained in F^{PV}. In fact there are, but this is not obvious, and since the isomorphism $\overline{F} \otimes_F R \to \overline{F}[G(F^{PV}/F)]$ will not be needed, we will not provide a proof of it here. Even without the Kolchin Theorem, one might ask about the Normal Basis Theorem; that is, whether there is a $G(F^{PV})$ module isomorphism $R \to F[G(F^{PV}/F)$? Assuming that the Kolchin Theorem holds for F^{PV} over F, this normal basis theorem comes down to asking if an infinite direct sum of copies of R that is isomorphic to an infinite direct sum of copies of $F[G(F^{PV}/F)]$ as $G(F^{PV}/F)$ modules implies that R is $G(F^{PV}/F)$ isomorphic to $F[G(F^{PV}/F)]$. Again since we will not need this, we will not provide a proof.

2.2 THE CATEGORY OF DIFFERENTIAL F MODULES

(The material in the first part of this section is based on "Universal covers and category theory in polynomial and differential Galois theory", Fields Inst. Commun., 43,American Mathematical Society, Providence, RI, 2004, 373–385.)

Although there are differential F modules of interest here which are not finite dimensional over F (R^{PV} being a good example), when we refer to the *category* of differential F modules we mean the category whose objects are differential F modules which are finite dimensional as F vector spaces, and whose morphisms are F linear differential maps. Having said that, our first task is to look at the contravariant . functor $V \mapsto \mathrm{Hom}_F^{\mathrm{diff}}(V, R^{PV})$ from differential F modules to abelian groups (actually to C vector spaces, since differential morphisms can be added and multiplied by constants). Every element of an F finite-dimensional differential module is a solution of a (monic, linear, homogeneous) differential equation, and hence so are their images under differential morphisms, from which it follows that $\mathrm{Hom}_F{}^{\mathrm{diff}}(V, F^{PV}) = \mathrm{Hom}_F{}^{\mathrm{diff}}(V, R^{PV})$.

The functor $V \mapsto \mathrm{Hom}_F^{\mathrm{diff}}(V, R^{PV})$ is right exact (carries surjections to injections). Our first major result is that it is also left exact, hence exact. This amounts to showing that if $V_0 \subset V$ is an inclusion of differential modules and if $f_0 : V_0 \to R^{PV}$ is a differential morphism then there exists a differential morphism $f : V \to R^{PV}$ such that f restricts to f_0 on V_0. We can do this step by step provided it can be done in the case that V is differentially generated over V_0 by a single element a (that is, V/V_0 is cyclic). So let $V_1 = \sum_i FD^i a$ be the cyclic submodule of V generated by a, so $V = V_0 + V_1$. Suppose there is a differential morphism $f_1 : V_1 \to R^{PV}$ such that $f_1(x) = f_0(x)$ for all $x \in V_0 \cap V_1$. Since $V \cong V_0 \oplus V_1/(\{(x, -x) | x \in V_0 \cap V_1\})$, $f : V \to R^{PV}$ by $v_0 + v_1 = f_0(v) + f_1(v_1)$ is well defined and extends f_0 to V. Thus proving that $V \mapsto \mathrm{Hom}_F^{\mathrm{diff}}(V, R^{PV})$ is left exact comes down to the following proposition:

Proposition 2.2 *Let V be a finite-dimensional cyclic differential F module and let $V_0 \subset V$ be a differential submodule. Let $f_0 : V_0 \to R^{PV}$ be* *a differential morphism. Then there is a differential morphism $f : V \to R^{PV}$ such that f restricted to V_0 is f_0.*

Before we begin the proof, we introduce some notation and make some observations. Let $\dim_F(V) = n$ and let a be a cyclic generator of V, so $V = \sum_{i=0}^{n-1} FD^i(a)$ with $D^n(a) = -\sum_{i=0}^{n-1} a_i D^i(a)$ and let $L = D^n + a_{n-1}D^{n-1} + \cdots + a_1 D + a_0$. Then the differential homomorphisms $f : V \to F^{PV}$ satisfy $L(f(a)) = 0$, and any element $x \in L^{-1}(0) \subset F^{PV}$ determines a homomorphism by sending $a \to x$. Since $L^{-1}(0) \subset R^{PV}$ these morphisms all have target R^{PV}. (Note that this implies that $\dim_C(\mathrm{Hom}_F^{\mathrm{diff}}(V, R^{PV})) \leq n$.) We know from Proposition 1.16 that V_0 is also cyclic say with generator b. Suppose that $\dim_F(V_0) = m$ so $b, D(b), \ldots, D^{m-1}(b)$ is a basis. Suppose $D^m(b) = \sum_{i=0}^{m-1} -b_i D^i(b)$, and let $L_0 = D^m + b_1 D^{m-1} + \cdots + b_1 D + b_0$. Then $L_0(b) = 0$, so $L_0(f_0(b)) = 0$. As with f, $f_0(b)$ determines f_0. Finally $b = \sum_{i=0}^{n-1} d_i D^i(a)$ for suitable $d_0 \ldots, d_{n-1}$ in F uniquely determined by b. Thus we can assume we have the following data: $a_0, \ldots, a_{n-1}$, the coefficients of L, from which we can define the cyclic differential module V. We also have $d_1, \ldots, d_{n-1}$ which determines b and hence the cyclic differential submodule V_0 and hence the differential operator L_0. We are given a morphism $V_0 \to R^{PV}$ which amounts to being given a solution $y \in F^{PV}$ to $L_0 = 0$. To find f, we need to find a solution $x \in F^{PV}$ to $L = 0$ so that $x = \sum_{i=0}^{n-1} d_i D^i(y)$. Of course $y \in L^{-1}(0) \subset T(E_L/F)$ and we have constructed E_L so that $T(E_L/F)$ is identified with R_L and in this latter $L^{-1}(0) = \sum_{i=1}^{n} Cx_{0i}$. Thus $x = c_1 x_{01} + \cdots + c_n x_{0n}$, $c_i \in C$; finding f thus comes down to solving for the c_i.

Although the notation of the preceding paragraph will be used in the proof of Proposition 2.2, unfortunately the author does not know any explicit formula for the above mentioned c_i's. Perhaps a clever reader can find one.

Proof. (Of Proposition 2.2). Let $\dim_F(V) = n$ and let $a \in V$ be a cyclic generator. Suppose $D^n(a) = -\sum_{i=0}^{n-1} a_i D^i(a)$, $a_i \in F$. We consider the symmetric algebra $S = S_F(V)$ and its subalgebra $S_0 = S_F(V_0)$. As we saw in Chapter 1, since V and V_0 are differential modules, S and S_0 are differential F algebras, with S_0 being a differential subalgebra of S. Since as a module $V = V_0 \oplus V_1$ for some vector subspace $V_1 \subset V$, as F algebras $S = S_0 \otimes_F S_F(V_1)$. Moreover, all these algebras are polynomial rings; so if $y_1, \ldots, y_k$ is a basis of V_1 then as F algebras $S = S_0[y_1, \ldots, y_k]$ is a polynomial ring over S_0. Let $X_i = D^i(a)$, $0 \leq i \leq n-1$. Then $\{X_0, \ldots, X_{n-1}\}$ is a basis of V so S is the n variable polynomial ring $F[X_0, \ldots, X_{n-1}]$ with the derivation D extending D_F and $D(X_i) = X_{i+1}$, $0 \leq i \leq n-2$, and $D(x_{n-1}) = -\sum_{i=0}^{n-1} a_i X_i$. The differential module homomorphism $f_0 : V_0 \to R^{PV}$ extends to a differential F algebra homomorphism $\phi_0 : S_0 \to R^{PV}$. Let Q_0, be the kernel of ϕ_0, let $\mathcal{K}$ be the quotient field of S_0/Q_0, and let K be the quotient field of the image $\phi_0(S_0)$. We have a differential field isomorphism $\sigma : \mathcal{K} \to K$. We have $\mathcal{K} \supset F$ and $K \supset F$ with σ the identity on F. Moreover, f_0 factors through σ. Since $K \subset F^{PV}$, $K_0 \supset F$ is a no new constants extension, and therefore so is $\mathcal{K} \supset F$. Now consider the ring $S_1 = S/Q_0S$. As an F algebra, S_1 is an integral domain, in fact a polynomial ring over S_0/Q_0: $S_1 = S_0/Q_0[y_1, \ldots, y_k]$. Using the embedding $S_0/Q_0 \to \mathcal{K}$ then we can form $S_2 = S_1[S_0/Q_0 - \{0\}]^{-1}$. As a ring S_2 is then an integral domain isomorphic to $\mathcal{K}[y_1,\ldots,y_k]$, so S_2 is an affine $\mathcal{K}$ algebra. As a differential $\mathcal{K}$ algebra, S_2 is differentially generated by the image x_0 of X_0. Finally, let Q be a maximal differential ideal of S_2 and let $\mathcal{K}_1$ be the quotient field of S_2/Q. Then $\mathcal{K}_1 \supset \mathcal{K}$ is a no new constants extension, where $\mathcal{K}_1$ generated as a differential field by $\mathcal{K}$ and the image $\overline{x}$ of x_0 which is also the image of a. By Lemma 2.1 there is a Picard–Vessiot extension $\mathcal{E} \supset \mathcal{K}$ for L with $\mathcal{K}_1 \subset \mathcal{E}$. Let E_0 be a Picard–Vessiot extension for L in F^{PV} and let $W = \{x \in E_0 | L(x) = 0$. Let E be the subfield of F^{PV} differentially generated by K and W. Then $E \supset K$ is a Picard–Vessiot extension for L. By Proposition 2.1 the differential field automorphism $\sigma : \mathcal{K} \to K$ lifts to a differential field isomorphism $\mathcal{E} \to E$ (note that since σ is the identity on F $L^\sigma = L$). The canonical map $V \to S_F(V) = S$ is differential, hence so is the composite $f : V \to S \to S_1 \to S_2 \to \mathcal{K}_1\mathcal{E} \to E$, where the last map is τ. This composite this restricts to $f_0 : V_0 \to S_F(V_0) = S_0 \to S_0/Q_0 \to S_0/Q_0[S_0/Q_0 - \{0\}]^{-1} \to K$ where the last map is σ. and thus f is the desired lift of f_0.

In the course of the proof of Proposition 2.2, we implicitly used the following, which we now make explicit:

Lemma 2.2 *Let* $L = D^n + a_{n-1}D^{n-1} + \cdots + a_1 D + a_0$ *be a differential operator over* F*. Let* V *be an* n *dimensional vector space over* F *with basis* $x_0,\ldots,x_{n-1}$ *and derivation* $D(x_i) = x_{i+1}$ *for* $i < n-1$ *and* $D(x_{n-1}) = -\sum_0^{n-1} a_i x_i$*.*

1. *If* $a \in F^{PV}$ *is such that* $L(a) = 0$ *then the* F *linear map* $V \to F^{PV}$ *by* $x_i \mapsto D^i(a)$ *is a morphism of differential modules.*

2. $Hom_F^{diff}(V, R^{PV}) \neq \{0\}$.

Proof. The first assertion is self-explanatory. For the second, as F^{PV} contains a Picard–Vessiot extension of F for L, $L^{-1}(0) \subset F^{PV}$ is of C dimension $n > 0$ so there is an $a \in F^{PV}$ with $a \neq 0$ and $L(a) = 0$. Then the corresponding differential map of the first assertion is non-zero, and since $a \in R^{PV}$ provides a non-zero element of the space of differential maps from V to R^{PV}.

The point of Proposition 2.2 (the left exactness) was to conclude exactness:

Corollary 2.1 $V \mapsto Hom_F^{diff}(V, R^{PV})$ *is an exact functor on* F *finite-dimensional differential* F *modules.*

On morphisms $\mathrm{Hom}_F^{\mathrm{diff}}(\cdot, R^{PV})$ acts by composition: if $T : V_1 \to V_2$ is a morphism of finite-dimensional differential F modules and $f : V_2 \to R^{PV}$ is a differential morphism, the $fT : V_1 \to R^{PV}$ is a differential morphism.

We note that $\mathrm{Hom}_F^{\mathrm{diff}}(V, R^{PV})$ is closed under addition, under scalar multiplication by constants, and also closed under composition with the group of differential automorphisms Π of R^{PV}. We are going to use the exactness of $V \to \mathrm{Hom}_F^{\mathrm{diff}}(V, R^{PV})$ to show that $\mathrm{Hom}_F^{\mathrm{diff}}(V, R^{PV})$ is of C dimension $\dim_F(V)$ and that the Π action on $\mathrm{Hom}_F^{\mathrm{diff}}(V, R^{PV})$ is rational.

To show that $\dim_C(\mathrm{Hom}_F^{\mathrm{diff}}(V, R^{PV})) = \dim_F(V)$, we observe that both sides are additive across exact sequences. We begin with the case that V is cyclic of positive dimension. Let a be a cyclic generator of V, so $V = \sum_{i=0}^{n-1} FD^i(a)$ with $D^n(a) = -\sum_{i=0}^{n-1} a_i D^i(a)$, and let $L = D^n + a_{n-1}D^{n-1} + \cdots + a_1 D + a_0$. Let $f : V \to R^{PV}$ be a differential homomorphism. The cyclicity of V says that $f(a)$ determines f; that f is differential implies that $L(f(a)) = 0$; and Lemma 2.2 implies that for any element $b \in L^{-1}(0)$ there is an f with $f(a) = b$. Thus $\mathrm{Hom}_F^{\mathrm{diff}}(V, R^{PV}) \to L^{-1}(0)$ by $f \mapsto f(a)$ is a bijection. It is also C linear, so we conclude $\dim_C(\mathrm{Hom}_F^{\mathrm{diff}}(V, R^{PV})) = \dim_F(V)$ when V is non-zero cyclic. The equality also holds when $V = \{0\}$. In general, we proceed by induction: if V is non-zero, it has a non-zero cyclic submodule V_0, and consider the exact sequence $0 \to V_0 \to V \to V/V_0 \to 0$. Because $\mathrm{Hom}_F^{\mathrm{diff}}(\cdot, R^{PV})$ is exact and both $\dim_C(\mathrm{Hom}_F^{\mathrm{diff}}(V_0, R^{PV})) = \dim_F(V_0)$ and $\dim_C(\mathrm{Hom}_F^{\mathrm{diff}}(V/V_0, R^{PV})) = \dim_F(V/V_0)$ by the induction assumption we have $\dim_C(\mathrm{Hom}_F^{\mathrm{diff}}(V, R^{PV})) = \dim_F(V)$.

To prove that $\mathrm{Hom}_F^{\mathrm{diff}}(V, R^{PV})$ is a rational Π module, we proceed as follows: let $a_1, \ldots, a_m$ be any spanning set for V. Suppose that a_i is a zero of the linear operator L_i. Let $W_i = L_i^{-1}(0) \subset F^{PV}$. We know that W_i is a finite-dimensional C vector space on which Π acts rationally. Let $W = W_1 \oplus \ldots W_m$; W is also a finite-dimensional rational Π module. Then $\mathrm{Hom}_F^{\mathrm{diff}}(V, R^{PV}) \to W$ by $f \mapsto (f(a_1), \ldots, f(a_m)$ is a Π equivariant injection of C spaces with Π action; call its image W_0. Since the action of Π on W is rational, so is the action on W_0 and hence on $\mathrm{Hom}_F^{\mathrm{diff}}(V, R^{PV})$. We sum up:

Theorem 2.1 *$V \mapsto \mathrm{Hom}_F^{\mathrm{diff}}(V, R^{PV})$ is an exact dimension preserving functor from F finite-dimensional differential F modules to C finite-dimensional rational Π modules.*

Let $T : V_1 \to V_2$ be a morphism of F differential modules. The induced map $\mathrm{Hom}_F^{\mathrm{diff}}(V_2, R^{PV}) \to \mathrm{Hom}_F^{\mathrm{diff}}(V_1, R^{PV})$, $f \mapsto fT$, is Π equivariant since the latter acts by composition on the left.

In the same way that differential maps from differential modules to R^{PV}, namely $\mathrm{Hom}_F^{\mathrm{diff}}(\cdot, R^{PV})$, provide a functor from differential modules to rational Π modules, Π equivariant maps from finite-dimesnional ration Π modules to R^{PV}, $\mathrm{Hom}_\Pi(\cdot, R^{PV})$, provides a functor to differential modules. Let W be a rational Π module, finite dimensional as a vector space over C. Let $\phi : W \to F^{PV}$ be Π equivariant. Because $\phi(W)$ is rational, it lies in R^{PV} (since it is the set of all elements of F^{PV} on which Π acts rationally), we consider $\phi : W \to R^{PV}$. The definition of rational action means that the corresponding representation map $\Pi \to GL(W)$ is a morphism of proalgebraic groups over C. Let H be its kernel. Then H is a Zariski closed normal subgroup of Π and the quotient proalgebraic group $G = \Pi/H \leq GL(W)$ is affine. We call such a subgroup *coaffine*. The point is that for any finite-dimensional rational Π modules, there is some coaffine closed normal subgroup of Π that acts trivially on it. We will adopt the notation that H is such a subgroup for W (not necessarily the kernel of the representation afforded by W (although of course contained in that kernel) and that $G = \Pi/H$ is the quotient. Since H acts trivially on W, W is a G module, still finite-dimensional and rational. Then any Π equivariant map $\phi : W \to R^{PV}$ has image in $(R^{PV})^H$. Then, since $(R^{PV})^H$ is all the H invariant elements of $(F^{PV})^H$, $(R^{PV})^H = T((R^{PV})^H/F)$, $\mathrm{Hom}_\Pi(W, R^{PV}) = \mathrm{Hom}_G(W, T((F^{PV})^H/F))$. Now, as previously noted, as a rational G module $T((F^{PV})^H/F)$ is isomorphic to $F[G] = F \otimes_C C[G]$ (here F is a trivial G module generally (the exception being when $F = C$) infinite dimensional over C. Thus we have $\mathrm{Hom}_\Pi(W, R^{PV})$ isomorphic to $\mathrm{Hom}_G(W, F[G])$. "Isomorphic" here means as F vector spaces (and hence as C vector spaces); we will add other structure later. Notice that G depends on W, namely that

$G = \Pi/H$ where H acts trivially on W. For any Π submodule $W_0 \subseteq W$ H also acts trivially on W_0, so it is also a G submodule. We are going to see that Π equivariant maps to R^{PV} from submodules of rational Π modules extend to Π equivariant maps of the ambient module, the Π analog of Proposition 2.2.

Proposition 2.3 *Let W be a C finite-dimensional rational Π module, $W_0 \subseteq W$ a Π submodule and $\phi_0 : W \to R^{PV}$ a Π equivariant C linear map. Then there is a Π equivariant C linear map $\phi : W \to R^{PV}$ such that $\phi|W_0 = \phi_0$.*

Proof. Let H be a coaffine closed normal subgroup of Π which acts trivially on W and let $G = \Pi/H$. As the discussion prior to the statement of the proposition explains, we can assume that ϕ_0 is a G morphism $W_0 \to F[G]$ which we want to lift to a G morphism $W \to F[G]$. Now as a G module $F[G] = F \otimes_C C[G]$, and F is a trivial G module. Write $F[G] = \varinjlim M_i \otimes_C C[G]$, where the M_i are the C finite dimensional C subspaces of F. Since W_0 is C finite dimensional, ϕ_0 has its image in $M_i \otimes C[G]$ for some i. In other words, we can assume ϕ_0 maps to $C[G]^{(n)}$ for some n, so $\phi_0 = (\psi_{01}, \ldots \phi_{0n})$ for some G morphism $\psi_{0j} : W_0 \to C[G]$. To find ϕ, it will suffice to find $\psi_i : W \to C[G]$ such that $\psi|W_0 = \psi_{0i}$. But now this is standard and follows from the fact that for any C finite-dimensional national G module N that $\mathrm{Hom}_G(N, C[G]) \to \mathrm{Hom}_C(N, C)$ by $f \mapsto f(e)$ (e is the identity of G) is a C vector space isomorphism (the inverse uses the $C[G]$ comodule structure $\gamma_N : N \to N \otimes C[G]$ to send $h \in \mathrm{Hom}_C(N, C)$ to $(h \otimes 1) \circ \gamma_N : N \to C[G]$). Then the surjection $\mathrm{Hom}_C(W, C) \to \mathrm{Hom}_C(W_0, C)$ shows that $\mathrm{Hom}_G(W, C[G]) \to \mathrm{Hom}_G(W_0, C[G])$ is surjective and hence there is a ψ_i lifting ψ_{0i}.

We note that since R^{PV} is an F vector space, so is $\text{Hom}_\pi(W, R^{PV})$. Two of the observations used in the proof of Proposition 2.3 imply the following F dimension formula for that vector space:

Proposition 2.4 *Let W be a C finite-dimensional rational Π module. Then*

$$dim_F(Hom_\Pi(W, R^{PV})) = dim_C(W).$$

Proof. Let H be a closed coaffine subgroup of Π acting trivially on W, and let $G = G/H$. As in the proof of Proposition 2.3, we can identify $\text{Hom}_\Pi(W, R^{PV})$ with $\text{Hom}_G(W, F[G])$. Write $F[G] = \varinjlim M_i \otimes_C C[G]$, where the M_i are the C finite dimensional C subspaces of F. Since W is C finite dimensional $\text{Hom}_G(W, \varinjlim M_i \otimes C[G])$ can be identified with $\varinjlim M_i \otimes \text{Hom}_G(W, C[G])$ so $F \otimes_C \text{Hom}_G(W, C[G]) \to \text{Hom}_G(W, F[G])$ by $x \otimes f \mapsto (w \mapsto xf(w)$ is a bijection. Since it is also F linear, it is an isomorphism of F vector spaces. In Proposition 2.3 we recalled that $\text{Hom}_G(W, C[G])$ is C linearly isomorphic to $\text{Hom}_C(W, C)$. Since W and $\text{Hom}_C(W, C)$ have the same C dimension, W and $\text{Hom}_G(W, C[G])$ have the same C dimension, which is equal to the F dimension of $F \otimes_C \text{Hom}_G(W, C[G])$ which thus has the same F dimension as $\text{Hom}_G(W, F[G])$, proving the proposition.

We have already remarked that if W is a C finite-dimensional rational Π module $\text{Hom}_\Pi(W, R^{PV})$ is an F vector space, which by Proposition 2.4 is F finite dimensional. It is, in fact, a differential module: if $\phi : W \to R^{PV}$ is C linear and G equivariant, define $D(\phi)$ by $D(\phi)(w) = D(\phi(w))$. Then for $a \in F$ and $w \in W$, $D(a\phi)(w) = D(a\phi(w) = D(a)\phi(w) + aD(\phi(w))$ so $D(a\phi) = D(a)\phi + aD(\phi)$. If $a \in C$, this means $D(\phi)$ is C linear. Moreover

$D(g\phi)(w) = D(g\phi(w)) = gD(\phi(w))$ so $D(\phi)$ is G equivariant. Thus $\phi \mapsto D(\phi)$ makes $\mathrm{Hom}_\Pi(W, R^{PV})$ a differential F module. Combining this observation with Propositions 2.3 and 2.4 thus gives the following analog of Theorem 2.1.

Theorem 2.2 *$W \mapsto Hom_\Pi(W, R^{PV})$ is an exact dimension preserving functor from C finite-dimensional rational Π modules to F finite-dimensional differential modules.*

On morphisms $\mathrm{Hom}_\Pi(\cdot, R^{PV})$ acts by composition: if $\tau : W_1 \to W_2$ is a morphism of C finite-dimensional rational Π modules and if $\phi : W_2 \to R^{PV}$ is Π equvariant and C linear then so is $\phi\tau$. The map $\phi \mapsto \phi\tau$ is differential because the differential structure is given by left composition with D.

Next we show that the functors of Theorems 2.1 and 2.2 are inverses of each other. We will do this via a sort of double dual argument. We consider

$$V \to \mathrm{Hom}_\Pi(\mathrm{Hom}_F^{\mathrm{diff}}(V, R^{PV}), R^{PV}), v \mapsto \hat{v}, \text{ where } \hat{v}(f) = f(v)$$

$$W \to \mathrm{Hom}_F^{\mathrm{diff}}(\mathrm{Hom}_\Pi(W, R^{PV}), R^{PV}), w \mapsto \hat{w} \text{ where } \hat{w}(\phi) = \phi(w)$$

We will show both maps to be isomorphisms. We begin by showing injectivity, which comes down to the following lemma:

Lemma 2.3 *Let V be a finite-dimensional differential module and $v \in V$, $v \neq 0$. Then there is a differential morphism $f : V \to R^{PV}$ with $f(v) \neq 0$.*

Let W be a finite-dimensional Π module and $w \in W$, $w \neq 0$. Then there is a Π morphism $\phi : W \to R^{PV}$ with $\phi(w) \neq 0$.

Proof. Let V_0 be the cyclic submodule of V generated by v, let $v, D(v), \ldots, D^{n-1}$ be a basis of V_0, and let $D^n(v) = -\sum_0^{n-1} a_i D^i(v)$. Let $L = D^n + a_{n-1}D^{n-1} + \cdots + a_1 D + a_0$. Let $y \in R^{PV}$, $y \neq 0$ be a solution of $L = 0$. Define a differential homomorphism $f_0 : V_0 \to R^{PV}$ by $v \mapsto y$. Note that $f_0(v) \neq 0$. Let $f : V \to R^{PV}$ be any differential morphism extending f_0. Then $f(v) \neq 0$. Let H be a closed coaffine normal subgroup of Π acting trivially on W. Let $G = \Pi/H$. We saw above in the proof of Proposition 2.3 that $\mathrm{Hom}_\Pi(W, R^{PV})$ can be identified with $\mathrm{Hom}_G(W, F[G])$ and then with $F \otimes_C \mathrm{Hom}_G(W, C[G])$. In particular, it suffices to find an element of $\mathrm{Hom}_G(W, C[G])$ which doesn't vanish on w. Since $\mathrm{Hom}_G(W, C[G]) \to \mathrm{Hom}_C(W, C)$ by composition with evaluation at the identity of G is an isomorphism, it suffices to find an element of $\mathrm{Hom}_C(W, C)$ which doesn't vanish on w. Such functionals always exist.

Lemma 2.3 shows that both of our double dual maps are injective. Because both double dualities preserve dimensions, these are injections between vector spaces of the same dimension; this proves they both are isomorphisms.

Theorem 2.3 *Let V be an F finite-dimensional differential F module then*

$$V \to Hom_\Pi(Hom_F^{diff}(V, R^{PV}), R^{PV}), v \mapsto \hat{v}, \text{ where } \hat{v}(f) = f(v)$$

is an isomorphism.

Let W be a C finite-dimensional rational Π module then

$$W \to Hom_F^{diff}(Hom_\Pi(W, R^{PV}), R^{PV}), w \mapsto \hat{w} \text{ where } \hat{w}(\phi) = \phi(w)$$

is an isomorphism.

Theorem 2.3 shows that the two exact contravariant functors $\mathrm{Hom}_F^{\mathrm{diff}}(\cdot, R^{PV})$ and $\mathrm{Hom}_\Pi(\cdot, R^{PV})$ are an inverse pair of equivalences between the category of F finite-dimensional F differential modules and the category of C finite-dimensional rational Π modules. Both categories are additive, and the functors carry the coproduct in each to the other: this is a consequence of the vector space facts that $\mathrm{Hom}_F(V_1, R^{PV}) \oplus \mathrm{Hom}_F(V_2, R^{PV}) \cong \mathrm{Hom}_F(V_1 \oplus V_2, R^{PV})$ (respectively $\mathrm{Hom}_C(W_1, R^{PV}) \oplus \mathrm{Hom}_C(W_2, R^{PV}) \cong \mathrm{Hom}_C(W_1 \oplus W_2, R^{PV})$), the observation that $\mathrm{Hom}_F^{\mathrm{diff}}(\cdot,\cdot) = \mathrm{Hom}_F(\cdot,\cdot)^D$ (respectively $\mathrm{Hom}_\Pi(\cdot,\cdot) = \mathrm{Hom}_C(\cdot, c)^{\cdot\Pi})$ and that $(\cdot)^D$ (respectively $(\cdot)^\Pi$) distributes over direct sums.

Although both categories also have tensor products, and we will eventually see that both the functors carry the tensor product in one category to the tensor product in the other, this requires more than the argument for the coproduct. Although we will see that the first part of the argument for coproducts also works for tensor products (the isomorphisms $\mathrm{Hom}_F(V_1, R^{PV}) \otimes_F \mathrm{Hom}_F(V_2, R^{PV}) \cong \mathrm{Hom}_F(V_1 \otimes_F V_2, R^{PV})$ and $\mathrm{Hom}_C(W_1, R^{PV}) \otimes_C \mathrm{Hom}_C(W_2, R^{PV}) \cong \mathrm{Hom}_C(W_1 \otimes_C W_2, R^{PV})$), which use the algebra structure of R^{PV}), neither $(\cdot)^D$ nor $(\cdot)^\Pi$ distribute over tensor product.

It turns out that what is needed to establish the preservation of tensor product is an analysis of differential modules which are the direct sums, as differential modules, of copies of the differential base field. We will return to the tensor product after an analysis of such.

2.3 TRIVIAL MODULES AND SPLITTING

When speaking of rational Π modules W, the module is called *trivial* if Π acts trivially on it; that is, $\sigma(w) = w$ for all $w \in W$ and $\sigma \in \Pi$. It is thus consistent, in light of Theorem 2.3, to refer to a finite-dimensional differential F module as *trivial* provided $\mathrm{Hom}_F^{\mathrm{diff}}(V, R^{PV})$ is a trivial Π module. As we will soon see, this implies that $V \cong F^{(n)}$ with coordinatewise derivation. Thus the modules $F^{(n)}$ (and modules differentially isomorphic to them) are called *trivial* differential modules.

Let $e_i \in F^{(n)}$ be the n-tuple with 1 in position i and 0 elsewhere, the standard basis. Let $p_i : F^{(n)} \to F$ be $p_i((a_1, \ldots, a_n)) = a_i$, the dual basis to the standard basis. Because $F^{(n)}$ has the coordiantewise derivation, p_i is differential as well as F linear, so $p_i \in \mathrm{Hom}_F^{\mathrm{diff}}(F^{(n)}, F)$. The p_i are linearly independent (over F and hence over C) and therefore form a C basis of $\mathrm{Hom}_F^{\mathrm{diff}}(F^{(n)}, F)$, since we know this latter has the same C dimension as the F dimension of $F^{(n)}$.

Of course, usually not all differential F modules V are trivial. It map happen that for a differential extension field $K \supset F$ that $K \otimes_F V$ is a trivial differential K module. When this happens, we say that K *splits* V and that K is a splitting field for V. We will see that splitting fields always exist, and that in the case $K \supset F^{PV}$ that they are unique.

We begin with an application of Theorem 2.3: we consider the case where V is a finite-dimensional differential module such that Π acts trivially on $\mathrm{Hom}_F^{\mathrm{diff}}(V, R^{PV})$. This means that as Π modules $\mathrm{Hom}_F^{\mathrm{diff}}(V, R^{PV})$ and $C^{(n)}$, the latter with trivial Π action, are Π isomorphic. Applying $\mathrm{Hom}_\Pi(\cdot, R^{PV})$ (which as we noted carries direct sums to direct sums) and Theorem 2.3 to this isomorphism then gives a differential module isomorphism $V \to \mathrm{Hom}_\Pi(C, R^{PV})^{(n)}$. On the other hand, it is clear that $\mathrm{Hom}_\Pi(C, R^{PV}) \doteq F$ as a differential module, so $V \cong F^{(n)}$, where the direct sum is differential componentwise. Now what does it mean for Π to act trivially on $\mathrm{Hom}_F^{\mathrm{diff}}(V, R^{PV})$? If there is a differential morphism $f : V \to R^{PV}$ and $v \in V$ such that $f(v) \notin F$, then there is a $\sigma \in \Pi$ such that $\sigma(f(v)) \neq f(v)$, and hence $\sigma(f) \neq f$ so Π would not be acting trivially on $\mathrm{Hom}_F^{\mathrm{diff}}(V, R^{PV})$. Thus π acts trivially if and only if $\mathrm{Hom}_F^{\mathrm{diff}}(V, R^{PV}) = \mathrm{Hom}_F^{\mathrm{diff}}(V, F)$. Since $\mathrm{Hom}_F^{\mathrm{diff}}(V, R^{PV}) = \mathrm{Hom}_F^{\mathrm{diff}}(V, F^{PV})$ we also have that Π acts trivially if and only if $\mathrm{Hom}_F^{\mathrm{diff}}(V, F^{PV}) = \mathrm{Hom}_F^{\mathrm{diff}}(V, F)$ We already saw that when Π acts trivially $V \cong F^{(n)}$. If, on the other hand, we have such an isomorphism then first, by preservation of direct sums, $\mathrm{Hom}_F^{\mathrm{diff}}(F^{(n)}, F^{PV})$ is isomorphic to $\mathrm{Hom}_F^{\mathrm{diff}}(F, F^{PV})^{(n)}$ and then since any differential homomorphism $f : F \to F^{PV}$ must send 1 to an element of C (since $D(1) = 0$ implies $D(f(1)) = 0$) and then linearity shows $f(F) = Ff(1) \subseteq F$. Thus.$\mathrm{Hom}_F^{\mathrm{diff}}(F, F^{PV}) = \mathrm{Hom}_F^{\mathrm{diff}}(F, F)$. We conclude that Π acts trivailly on $\mathrm{Hom}_F^{\mathrm{diff}}(F^{(n)}, F^{PV})$. We record these observations, which ultimately depend on Theorem 2.3 and hence are a corollary:

Corollary 2.2 *A finite-dimensional F differential module V is isomorphic to $F^{(n)}$ with coordinatewise derivation if and only if*

$$dim_C(Hom_F^{diff}(V, F)) = dim_F(V).$$

Proof. Since $\mathrm{Hom}_F^{\mathrm{diff}}(V, F)$ is a subspace of $\mathrm{Hom}_F(V, F^{PV})$, and we know the latter has C dimension equal to the F dimension of V, the two spaces coincide if and only if the two dimensions are equal. Since we also know that the two spaces coincide if and only if V is isomorphic to $F^{(n)}$, the corollary follows.

Notice that Corollary 2.2 mentions only the field F; the field F^{PV} and the ring R^{PV} are not mentioned.

Let E be a no-new-constants differential extension field of F. This makes C the field of constants for E as well. Let V be a differential F module of differential dimension n. Consider $E \otimes_F V$ as an E vector space (which is also of dimension n). Then $\mathrm{Hom}_E(E \otimes_F V, E) \to \mathrm{Hom}_F(V, E)$ by $f \mapsto (v \mapsto f(1 \otimes v))$ is bijective, with inverse $h \mapsto (e \otimes v \mapsto eh(v))$. If f (respectively h) is differential then so are their respective images, so $\mathrm{Hom}_E^{\mathrm{diff}}(E \otimes_F V, E) \to \mathrm{Hom}_F^{\mathrm{diff}}(V, E)$ is bijective. Combining this observation with Corollary 2.2 then yields the following important fact.

Theorem 2.4 *Let V be a differential F module of F dimension n. Let E be a no new constants extension of F. If $\dim_C(\mathrm{Hom}_F^{\mathrm{diff}}(V, E)) = n$ then $E \otimes_F V$ is isomorphic to $E^{(n)}$ (coordinatewise derivation) as differential E modules. If E is a subfield of F^{PV} then $\dim_C(\mathrm{Hom}_F^{\mathrm{diff}}(V, E)) = n$ if and only if $\mathrm{Hom}_F^{\mathrm{diff}}(V, E) = \mathrm{Hom}_F^{\mathrm{diff}}(V, F^{PV})$.*

Proof. Since $\mathrm{Hom}_E^{\mathrm{diff}}(E \otimes_F V, E) \cong \mathrm{Hom}_F^{\mathrm{diff}}(V, E)$ both have the same C dimension. By hypothesis, the C dimension of $\mathrm{Hom}_F^{\mathrm{diff}}(V, E)$ is n, so the C dimension of $\mathrm{Hom}_E^{\mathrm{diff}}(E \otimes_F V, E)$ is also n. Now apply Corollary 2.2 for the case of E and $E \otimes_F V$.

Proof. The final assertion is a consequence of the facts that $\mathrm{Hom}_F^{\mathrm{diff}}(V, F^{PV}) = \mathrm{Hom}_F^{\mathrm{diff}}(V, R^{PV})$ and $\dim_C(\mathrm{Hom}_F^{\mathrm{diff}}(V, R^{PV}) = \dim_F(V)$.

Theorem 2.4 has a sort of converse.

Proposition 2.5 *Let V be a finite-dimensional differential F module. Suppose $K \supseteq F$ is a differential subfield of F^{PV} such that $K \otimes_F V$ is* *isomorphic to* $K^{(n)}$ *as a differential* K *module. Then* $Hom_F^{diff}(V, F^{PV}) = Hom_F^{diff}(V, K)$*, so* $dim_C(Hom_F^{diff}(V, K)) = n$*.*

Proof. The constants of K are C. We can construct K^{PV} so that $F^{PV} \subseteq K^{PV}$. The C vector space $\text{Hom}_K^{\text{diff}}(K^{(n)}, K^{PV})$ has dimension n. Let $p_i : K^{(n)} \to K$ be projection on the ith summand. Each p_i is a morphism of differential E modules, and together they are linearly independent over C so they form a C basis of $\text{Hom}_K^{\text{diff}}(K^{(n)}, K^{PV})$. Let $f : V \to F^{PV}$ be a differential morphism. Following f by the inclusion gives the F differential morphism $f : V \to K^{PV}$ which in turn gives the K differential morphism $1 \otimes f : K \otimes V \to K^{PV}$ by $(1 \otimes f)(k \otimes v) = kf(v)$. Thus $1 \otimes f = \sum c_i p_i$ for some c_i's in C. So for $v \in V$ $f(v) = (1 \otimes f)(1 \otimes v) = \sum c_i p_i(v) \in K$. That is, $f(V) \subset K$. So $\text{Hom}_F^{\text{diff}}(V, F^{PV}) = \text{Hom}_F^{\text{diff}}(V, K)$ from which the dimension assertion of the proposition follows.

In other terminology, Theorem 2.4 and Proposition 2.5 give criteria for a differential extension field of F to be a splitting field for V.

There are, of course, many differential subfields E of F^{PV} containing F for which $\text{Hom}_F^{\text{diff}}(V, E) = \text{Hom}_F^{\text{diff}}(V, F^{PV})$, for example F^{PV} itself. We turn our attention to he smallest such. Any such field E must contain the image of any differential F morphism $f : V \to F^{PV}$. So we start with fields which contain, as any such field must,

$$\mathcal{I}_V = \{f(v) | f \in \text{Hom}_F^{\text{diff}}(V, F^{PV}), v \in V\}$$

Note that $\mathcal{I}_V \subset R^{PV}$.

If $\sigma \in \Pi$, and $f \in \text{Hom}_F^{\text{diff}}(V, F^{PV})$ then $\sigma(f(v)) = \sigma(f)(v)$ so $\mathcal{I}_V$ is Π stable, and $D(f(v) = f(D(v))$ so $\mathcal{I}_V$ is D stable. The same is true for its F span $F\mathcal{I}_V$.

Let E_V be the subfield of F^{PV} generated over F by $\mathcal{I}_V$. Since $\mathcal{I}_V$ is Π stable so is E_V and since $\mathcal{I}_V$ is D stable E_V is a differential field. We have by construction that $\text{Hom}_F^{\text{diff}}(V, E_V) = \text{Hom}_F^{\text{diff}}(V, F^{PV})$ and further that if $\text{Hom}_F^{\text{diff}}(V, E) = \text{Hom}_F^{\text{diff}}(V, F^{PV})$ then $E_V \subseteq E$.

Further, E_V is finitely generated over F as a field:

We can also describe $F\mathcal{I}_V$ as the image of the map

$$\Phi : \text{Hom}_F^{\text{diff}}(V, F^{PV}) \otimes_C V \to F^{PV}, f \otimes v \mapsto f(v)$$

Now $\text{Hom}_F^{\text{diff}}(V, F^{PV}) \otimes_C V$ is a Π module by $\sigma(f \otimes v) = \sigma(f) \otimes v$ (action on the first factor and an F differential module by $a(f \otimes v) = f \otimes av$ and $D(f \otimes v) = f \otimes D(v)$ (action on the second factor) and Φ is then F linear, differential, and Π equivariant. As an F vector space $\text{Hom}_F^{\text{diff}}(V, F^{PV}) \otimes_C V$ is n^2 dimensional; if $f_1, \ldots, f_n$ is any C basis of $\text{Hom}_F^{\text{diff}}(V, F^{PV})$ and $v_1, \ldots, v_n$ is any F basis of V then $f_i \otimes v_j$ is a basis of $\text{Hom}_F^{\text{diff}}(V, F^{PV}) \otimes_C V$, and $\{f_i(v_j) | 1 \leq i, j \leq n\}$ is a spanning set for $F\mathcal{I}_V$.

We collect these observations as a theorem:

Theorem 2.5 *Let V be an F finite-dimensional differential module. Let E_V be the subfield of F^{PV} generated over F by $\{f(v) | f \in Hom_F^{diff}(V, F^{PV}), v \in V\}$. Then E_V is a differential subfield of F^{PV} stable under Π. E_V is a splitting field for V, and is contained in any splitting field for V contained in F^{PV}. If $f_1, \ldots, f_n$ is any C basis of $Hom_F^{diff}(V, F^{PV})$ and $v_1, \ldots, v_n$ is any F basis of V then $\{f_i(v_j) | 1 \leq i, j \leq n\}$ generates E_V as a field over F. In particular, E_V is a Picard–Vessiot extension of F.*

Proof. All of the assertions of the theorem except the final two sentences have already been observed. For them, let $W = \sum_{i,j} C f_i(v_j)$.

Note that W is Π stable and C finite dimensional. We claim that $FW = \sum_{i,j} Ff_i(V_j) = F\mathcal{I}_V$. If $f \in \mathrm{Hom}_F^{\mathrm{diff}}(V, F^{PV})$ and $v \in V$ there are $c_1, \ldots, c_n \in C$ and $a_1, \ldots, a_n \in F$ such that $f = \sum_i c_i f_i$ and $v = \sum a_j v_j$ so that $f(v) = \sum_{i,j} c_i f_i(a_j v_j) = \sum_{i,j} c_i a_j f_i(v_j)$ so $f(v) \in FW$. Thus $\mathcal{I}_V \subset FW$ and hence $F\mathcal{I}_V \subseteq FW$. The reverse inclusion is obvious and hence $FW = F\mathcal{I}_V$. Thus they both generate the same field, namely E_V. This proves the assertion of the penultimate sentence of the statement of the theorem. We already know that E_V is a differential subfield of F^{PV}, hence the fixed field of the subgroup $H = \mathrm{Aut}^{\mathrm{diff}}(F^{PV}/E_V)$ and that H is a normal subgroup of Π because E_V is Π stable. Now that we have that E_V is finitely generated over F, we know that H is coaffine, which established the assertion made in the final sentence of the statement.

Let V be an F finite-dimensional differential module. Since E_V splits V, so does every differential field E between E_V and F^{PV}, including F^{PV} itself. The converse is also true:

Proposition 2.6 *F^{PV} is a splitting field for every F finite-dimensional differential module. Conversely, suppose E is a differential subfield of F^{PV} which contains F and which is a splitting field for every F finite-dimensional differential module. Then $E = F^{PV}$.*

Proof. We have already observed the first assertion. For the second, assume it is false and let $E_0 \subseteq F^{PV}$ be a Picard–Vessiot extension of F not contained in E; such exists since F^{PV} is a union of Picard–Vessiot extensions of F. Let L be a monic, linear, homogeneous differential operator over F for which E_0 is a Picard–Vessiot extension. Let $L = D^n + a_{n-1}D^{n-1} + \cdots + a_1 D + a_0$. Let V be an n dimensional vector space over F with basis $x_0, \ldots, x_{n-1}$ and derivation $D(x_i) = x_{i+1}$ for

$i < n-1$ and $D(x_{n-1}) = -\sum_0^{n-1} a_i x_i$. As noted in Lemma 2.2 if $a \in F^{PV}$ is such that $L(a) = 0$ then the F linear map $V \to F^{PV}$ by $x_i \mapsto D^i(a)$ is a morphism of differential modules. Since $\{a \in F^{PV} | L(a) = 0\}$ is contained in E_0 and generates it, there must be some $a \in F^{PV}$ with $L(a) = 0$, $a \notin E$, and hence a differential morphism $f : V \to F^{PV}$ with $f(x) = a \notin E$. But since E is a splitting field for V by Proposition 2.5, every differential homomorphism from V to F^{PV} has image contained in E. This contradiction shows that $E = F^{PV}$.

If $W \subset F^{PV}$ is any C finite-dimensional subspace stable under Π then (as shown in ULECT7) there is a monic linear homogeneous differential operator $L(Y)$ of order $m = \dim_C(W)$ with coefficients in F such that W is the full set of solutions of L in F^{PV}. (If $w_1, \ldots, w_m$ is a C basis of W, the ratio of Wronskians $L(Y) = w(w_1, \ldots, w_m)^{-1} w(Y, w_1, \ldots, w_m)$ has this property.) Thus the differential subfield of F^{PV} generated by W will be a Picard–Vessiot extension of F for L. If W is the space $\sum_{i,j} C f_i(v_j)$ from the proof of Theorem 2.5 then $\dim_C(W) \leq n^2$.

To illustrate, we consider the case $V = F^{(n)}$ with coordinate-wise derivation. We can take the standard basis $e_1, \ldots, e_n$ as the basis of V and its dual basis $p_1, \ldots, p_n$ (recall p_i is projection on the ith coordinate) as a basis for $\text{Hom}_F^{\text{diff}}(F^{(n)}, F)$. Then $p_i(e_j) = \delta_{ij}$. The subspace "W" is then just C, which has C dimension 1 and the subfield of F^{PV} generated by the $p_i(e_j)$ is F, which is, of course, a splitting field for the already split $F^{(n)}$. In this case we have $\dim_C(W) = 1$.

The definition of the splitting field E_V and the generating set

$$\{f(v) | f \in \text{Hom}_F^{\text{diff}}(V, F^{PV}), v \in V\}$$

for $E_V$of Theorem 2.5 depend only on V and not on any basis of V or $\mathrm{Hom}_F^{\mathrm{diff}}(V, F^{PV})$, while of course the finite generating set $\{f_i(v_j)|1 \leq i,j \leq n\}$ depends on both. We want to consider this latter set to see explicitly how E_V splits V.

We begin by considering the $n \times n$ matrix $X = [f_i(v_j)]$ whose entries generate the field E_V as a differential field over F.

To take into consideration the dependence of X on the choice of bases, we introduce the following notation:

Let $\mathcal{F} = \{f_1, \ldots, f_n\}$ be a basis of $\mathrm{Hom}_F^{\mathrm{diff}}(V, F^{PV})$ and $\mathcal{B} = \{v_1, \ldots, v_n\}$ be a basis of V then

$$X(\mathcal{F}, \mathcal{B}) = [f_i(v_j)] \in M_n(F^{PV}).$$

We want to describe what happens to X when we change bases. Suppose $\{h_1, \ldots, h_n\}$ is another basis of $\mathrm{Hom}_F^{\mathrm{diff}}(V, F^{PV})$. Then there are elements c_{ij}, $1 \leq i,j \leq n$ such that $h_i = \sum_{j=1}^n c_{ji} f_j$ with $[c_{ij}] \in GL_n(C)$. We write this as $\{h_1, \ldots, h_n\} = \mathcal{F}[c_{ij}]$. Similarly, if $\{y_1, \ldots, y_n\}$ is another basis of V then there is $[b_{ij}] \in GL_n(F)$ such that $y_i = \sum_{j=1}^n b_{ij} v_j$. We write this as $\{y_1, \ldots, y_n\} = [b_{ij}]\mathcal{B}$. Then we have the formula

$$X(\mathcal{F}[c_{ij}], [b_{ij}]\mathcal{B}) = [b_{ij}] X(\mathcal{F}, \mathcal{B})[c_{ij}].$$

Next we want to describe what happens when we extend scalars. Let $E \supseteq F$ be a subfield of F^{PV}. For each i we have $1 \otimes f_i : E \otimes_F V \to F^{PV}$ by $(1 \otimes f_i)(x \otimes v) = x f_i(v)$ is an E linear differential homomorphism. Also $\{1 \otimes v_1, \ldots, 1 \otimes v_n\}$ is a basis of $E \otimes V$. And we have matrix equality

$$[(1 \otimes f_i)(1 \otimes v_j)] = X(\mathcal{F}, \mathcal{B}) \text{ in } M_n(E).$$

Now suppose $E \supseteq E_V$. Then $E \otimes V$ is differentially isomorphic to $E^{(n)}$ with coordinate-wise derivation. Suppose that under this isomorphism $h_i \in \mathrm{Hom}_E^{\mathrm{diff}}(E \otimes V, E)$ corresponds to the coordinate projection p_i and $y_i \in E \otimes V$ corresponds to the standard basis vector e_i. Then $[h_i(y_j)] = [p_i(e_j)] = [\delta_{ij}]$ is the identity matrix. Further, there are matrices $[c_{ij}] \in GL_n(C)$ and $[b_{ij}] \in GL_n(E)$ such that $\{h_1, \dots, h_n\} = \{f_1, \dots, f_n\}[c_{ij}]$ and $\{y_1, \dots, y_n\} = [b_{ij}]\{1 \otimes v_1, \dots, 1 \otimes v_n\}$.Thus

$$[b_{ij}]X(\mathcal{F}, \mathcal{B})[c_{ij}] = [\delta_{ij}] = I_n.$$

Taking determinants, we see $\det(X(\mathcal{F}, \mathcal{B})) \neq 0$ so that $X(\mathcal{F}, \mathcal{B})$ is invertible. We record this:

Proposition 2.7 *Let V be an F finite-dimensional differential module. If $\{f_1, \dots, f_n\}$ is any C basis of $Hom_F^{diff}(V, F^{PV})$ and $\{v_1, \dots, v_n\}$ is any F basis of V then the $n \times n$ matrix $[f_i(v_j)]$ belongs to $GL_n(E_V)$.*

We retain the notations f_i and v_j of Proposition 2.7. Using the F vector space isomorphism $V \to F^{(n)}$ given by $v_i \mapsto e_i$, we have a differential module isomorphism $V \to (R^{(n)}, {}_A D)$: $A = [a_{ij}]$ is defined by $D(v_i) = \sum_j a_{ij} v_j$. Let $X = [f_i(v_j)]$. We want to apply Proposition 1.12 which expresses the fact that the f_i are differential in terms of the equality of the columns

$$\begin{bmatrix} D(f_i(v_1)) \\ D(f_i(v_2)) \\ \dots \\ D(f_i(v_n)) \end{bmatrix} = A \begin{bmatrix} f_i(v_1) \\ f_i(v_2) \\ \dots \\ f_i(v_n) \end{bmatrix}$$

namely that

$$(X^T)' = AX^T \text{ in } M_n(E_V).$$

We recall that by Corollary 1.8 that this means that (F^n, ∂_A) has a basis of constants, namely the columns of X^T, and that $(F^{(n)},_A \partial)$ (our V) has a basis of constants, namely the rows of $(X^T)^{-1}$. Since X^{-1} is the transposed matrix of cofactors of X divided by the determinant of X, so $(X^T)^{-1} = (X^{-1})^T$ is the matrix of cofactors of X divided by the determinant, and we ultimately have a formula for the basis of constants of $E_V \otimes_F V$.

Theorem 2.6 *Let V be an F finite-dimensional differential module of dimension n. Let $\{f_1, \ldots, f_n\}$ be a C basis of $Hom_F^{diff}(V, F^{PV})$ and let $\{v_1, \ldots, v_n\}$ be an F basis of V. Let X be the $n \times n$ matrix $[f_i(v_j)]$. Let g_{ij} be the i,j cofactor of X divided by the determinant of X, and $y_i = \sum_j g_{ij} \otimes v_i$, $1 \leq i \leq n$. Then $E_V \otimes_F V \to E_V^{(n)}$ by $y_i \mapsto e_i$ is a differential isomorphism.*

To illustrate, we consider the case where V is a cyclic module of dimension n generated by $v \in V$ with $D^n(v) = -\sum_{i=0}^{n-1} a_i D^i(v)$. Let M be the differential operator $Y^{(n)} + a_{n-1}Y^{(n-1)} + \ldots a_1 Y + a_0$. As a basis of V we can take $\{D^i(v) | 0 \leq i \leq n-1\}$. For any $h \in F^{PV}$ with $M(h) = 0$, we can define an F linear map $V \to F^{PV}$ by $\sum_0^{n-1} b_i D^i(a) \mapsto \sum b_i D^i(h)$; such a map is F linear and differential. Since M has a complete set of solutions in F^{PV}, $\dim_C(M^{-1}(0)) = n$; let $h_1, \ldots, h_n$ be a C basis of $M^{-1}(0)$. Then the maps $f_i : V \to F^{PV}$ defined by $v \mapsto h_i$ form a C basis of $\mathrm{Hom}_F^{\mathrm{diff}}(V, F)$. Here we have $f_i(D^j(v)) = D^j(f_i(v)) = D^j(h_i)$. Let $W_0 = \sum Ch_i$ $(W_0 = M^{-1}(0))$, so the differential field generated by W_0 is the Picard–Vessiot extension for M in F^{PV}. Then $W = \sum_0^{n-1} D^i(W_0)$

so that the differential field E_V generated by W is the differential field generated by W_0; in other words, the splitting field E_V for the cyclic module V associated to the differential operator M is the Picard–Vessiot extension for M.

We continue to retain the notation of this example, except writing v_{i+1} for the basis element $D^i(v)$. We describe the derivation on V using a "mixed matrix" notation. If $v \in V$ is expressed in the basis as $v = \sum x_i v_i$ then we write

$$v = \begin{bmatrix} x_1, & x_2, & \ldots, & x_n \end{bmatrix} \begin{bmatrix} v_1 \\ v_2 \\ \ldots \\ v_n \end{bmatrix}$$

Then

$$D(v) = D(\begin{bmatrix} x_1, & x_2, & \ldots, & x_n \end{bmatrix}) \begin{bmatrix} v_1 \\ v_2 \\ \ldots \\ v_n \end{bmatrix} + \begin{bmatrix} x_1, & x_2, & \ldots, & x_n \end{bmatrix} D(\begin{bmatrix} v_1 \\ v_2 \\ \ldots \\ v_n \end{bmatrix})$$

To expand these, we have first that

$$D(v) = D(\begin{bmatrix} x_1, & x_2, & \ldots, & x_n \end{bmatrix}) = \begin{bmatrix} D(x_1), & D(x_2), & \ldots, & D(x_n) \end{bmatrix}$$

then that

$$D(\begin{bmatrix} v_1 \\ v_2 \\ \ldots \\ v_n \end{bmatrix}) = \begin{bmatrix} v_2 \\ v_3 \\ \ldots \\ v_n \\ -a_0 v_1 - \cdots - a_{n-1} v_n \end{bmatrix}$$

$$= \begin{bmatrix} 0 & 1 & \dots & 0 \\ 0 & 0 & 1\dots & 0 \\ \dots & & & \\ 0 & 0 & \dots & 1 \\ -a_0 & -a_1 & \dots & -a_{n-1} \end{bmatrix} \begin{bmatrix} v_1 \\ v_2 \\ \dots \\ v_n \end{bmatrix}$$

thus $D(v) =$

$$(\begin{bmatrix} D(x_1), & D(x_2), & \dots, & D(x_n) \end{bmatrix} + \begin{bmatrix} 0 & 1 & \dots & 0 \\ 0 & 0 & 1\dots & 0 \\ \dots & & & \\ 0 & 0 & \dots & 1 \\ -a_0 & -a_1 & \dots & -a_{n-1} \end{bmatrix}) \begin{bmatrix} v_1 \\ v_2 \\ \dots \\ v_n \end{bmatrix}$$

We then use the basis $v_1, \dots, v_n$ to set up an isomorphism $V \to F^{(n)}$ as F vector spaces. Then the derivation $D = D_V$ on V becomes the derivation ${}_AD$ on $F^{(n)}$ where

$$A = \begin{bmatrix} 0 & 1 & \dots & 0 \\ 0 & 0 & 1\dots & 0 \\ \dots & & & \\ 0 & 0 & \dots & 1 \\ -a_0 & -a_1 & \dots & -a_{n-1} \end{bmatrix}$$

Using the basis $v_1, \dots, v_n$ we have constructed a differential isomorphism $V \to (F^{(n)}, {}_AD)$.

We already have a specified F basis $v_1, \dots, v_n$ for V. As a C basis for $\mathrm{Hom}_F^{\mathrm{diff}}(V, F^{PV})$, we can take the homomorphisms f_i defined by $f_i(v_1) = h_i$, so $f_i(v_j) = D^j(h_i)$. Then the matrix $X = [f_i(h_j)]$ is the matrix whose entry in row i and column j is $D^j(h_i)$ (the transposed Wronskian matrix):

$$X = \begin{bmatrix} h_1 & D(h_1) & \dots & D^{n-1}(h_1) \\ h_2 & D(h_2) & \dots & D^{n-1}(h_2) \\ \dots & & & \\ h_n & D(h_n) & \dots & D^{n-1}(h_n) \end{bmatrix} = W(h_1, \dots, h_n)^T$$

To compute $D(X)$, it is convenient to let X_i denote the jth column of X, Then $D(X_j) = X_{j+1}$ for $1 \leq j \leq n-1$ with $D(X_n) = -\sum_{j=0}^{n-1} a_j X_{j+1}$. Thus

$$D(X) = XA^T \text{ or } D(X^T) = AX^T$$

where X^T is the Wronskian matrix $W(h_1, \dots, h_n)$. It then follows from Theorem 2.6 that $y_i = \sum g_{ij} \otimes v_j$, where g_{ji} is the j,i cofactor of $W(h_1, \dots, h_n)$ divided by the Wronskian determinant $w(h_1, \dots, h_n)$.

For an easy concrete example, we take the field F to be $\mathbb{C}$ (the complex numbers), with the zero derivative. We will discuss $\mathbb{C}^{PV}$ further below; for now it suffices to know that it contains the field $\mathbb{C}(t)$, the field of complex rational functions in one variable, whose derivation is d/dt. For V we take the module associated with $Y'' = 0$. V has a basis v_1, v_2 with $D(v_1) = v_2$ and $D(v_2) = 0$. As Y'' has degree 2 and the elements 1 and t of $\mathbb{C}(t)$ are linearly independent over $\mathbb{C}$ and satisfy $1'' = 0$ and $t'' = 0$, they are a basis for the full set of solutions of $Y'' = 0$ in $\mathbb{C}^{PV}$. We have differential homomorphisms $f_1 : V \to \mathbb{C}(t)$ by $v_1 \mapsto t$ (and $v_2 \mapsto 1$) and $f_2 : V \to \mathbb{C}(t)$ by $v_1 \mapsto 1$ (and $v_2 \mapsto 0$). These are linearly independent over $\mathbb{C}$ since their images of v_1 are and hence form a basis of $\mathrm{Hom}_{\mathbb{C}}^{\mathrm{diff}}(V, \mathbb{C}^{PV})$. The field generated over $\mathbb{C}$ by $f_1(v_1) = t, f_1(v_2) = 1, f_2(v_1) = 1, f_2(v_2)$ is

$\mathbb{C}(t)$ so $E_V = \mathbb{C}(t)$. In this case we have

$$X = \begin{bmatrix} f_1(v_1) & f_1(v_2) \\ f_2(v_1) & f_2(v_2) \end{bmatrix} = \begin{bmatrix} t & 1 \\ 1 & 0 \end{bmatrix} \text{ and } X^T = \begin{bmatrix} t & 1 \\ 1 & 0 \end{bmatrix}$$

The matrix A in this case is

$$\begin{bmatrix} 0 & 1 \\ 0 & 0 \end{bmatrix}$$

and

$$D(X^T) = \begin{bmatrix} 1 & 0 \\ 0 & 0 \end{bmatrix} = AX^T = \begin{bmatrix} 0 & 1 \\ 0 & 0 \end{bmatrix} \begin{bmatrix} t & 1 \\ 1 & 0 \end{bmatrix}$$

The determinant of X is -1, and the matrix of cofactors of X is

$$\begin{bmatrix} 0 & -1 \\ -1 & t \end{bmatrix}.$$

The basis y_1, y_2 of $\mathbb{C}(t) \otimes V$ is then $y_1 = 0 \otimes v_1 + 1 \otimes v_2 = 1 \otimes v_2$ and $y_2 = 1 \otimes v_1 + -t \otimes v_2$. We have $D(y_1) = 1 \otimes D(v_2) = 0$ and $D(y_2) = 1 \otimes D(v_1) + D(-t) \otimes v_2 + -t \otimes D(v_2) = 1 \otimes v_2 - 1 \otimes v_2 + -t \otimes 0 = 0$ so y_1 and y_2 are constants. Also since $1 \otimes v_2 = y_1$ and $1 \otimes v_1 = y_2 + ty_1$ it is clear that y_1, y_2 is a basis.

2.4 APPLICATIONS OF CATEGORICAL EQUIVALENCE

As noted above both the categories of differential F modules and Π modules have tensor products ($V_1 \otimes_F V_2$ for differential modules and $W_1 \otimes_C W_2$ for Π modules). Now we show that both the functors of Theorem 2.3 carry the tensor product in one category to the tensor product in the other:

Proposition 2.8 *Let V_1 and V_2 be F finite-dimensional differential modules. Then*

$$Hom_F^{diff}(V_1, F^{PV}) \otimes_C Hom_F^{diff}(V_2, F^{PV}) \to Hom_F^{diff}(V_1 \otimes_F V_2, F^{PV}),$$

$$f_1 \otimes f_2 \mapsto (v_1 \otimes v_2 \mapsto f_1(v_1) f_2(v_2))$$

is an isomorphism.

Let W_1 and W_2 be C finite-dimensional rational Π modules. Then

$$Hom_\Pi(W_1, F^{PV}) \otimes_F Hom_\Pi(W_2, F^{PV}) \to Hom_\Pi(W_1 \otimes_C W_2, F^{PV}),$$

$$\phi_1 \otimes \phi_2 \mapsto (w_1 \otimes w_2 \mapsto \phi_1(w_1) \phi_2(w_2))$$

is an isomorphism.

We recall that $\mathrm{Hom}_F^{\mathrm{diff}}(V_i, F^{PV}) = \mathrm{Hom}_F^{\mathrm{diff}}(V_i, R^{PV})$ and $\mathrm{Hom}_\Pi (W_i, F^{PV}) = \mathrm{Hom}_\Pi(V_i, R^{PV})$. Proposition 2.8 is stated as it is to facilitate steps in the proof, which we now present.

Proof. (Of Proposition 2.8.) Let E be a differential subfield of F^{PV} containing F (both extremes are also allowed). Then for a differential F module V $\mathrm{Hom}_E((E \otimes_F V, F^{PV}) \to \mathrm{Hom}_F(V, F^{PV})$ by $f \mapsto (v \mapsto f(1 \otimes v))$ is, as previously noted, bijective, with inverse $h \mapsto (e \otimes v \mapsto eh(v))$. If f (respectively h) is differential then so are their respective images, so $\mathrm{Hom}_E^{\mathrm{diff}}(E \otimes_F V, F^{PV}) \to \mathrm{Hom}_F^{\mathrm{diff}}(V, F^{PV})$ is bijective. Suppose E is chosen so that $\mathrm{Hom}_F^{\mathrm{diff}}(V_i, F^{PV}) = \mathrm{Hom}_F^{\mathrm{diff}}(V_i, E)$, $i = 1, 2$. (Since $\mathrm{Hom}_F^{\mathrm{diff}}(V_i, F^{PV})$, $i = 1, 2$ is C finite dimensional, there is a Picard–Vessiot extension E with this property, although we don't need that here.) Apply the bijection with $V = V_i$ and $E = F^{PV}$. Then the first assertion of the Proposition amounts to verifying that

$$\mathrm{Hom}_E^{\mathrm{diff}}(E \otimes_F V_1, E) \otimes_C \mathrm{Hom}_E^{\mathrm{diff}}(E \otimes_F V_2, E)$$
$$\to \mathrm{Hom}_E^{\mathrm{diff}}((E \otimes_F V_1) \otimes_E (E \otimes_F V_2), E)$$

is bijective. Since $\mathrm{Hom}_F^{\mathrm{diff}}(V, E) = \mathrm{Hom}_F^{\mathrm{diff}}(V, F^{PV})$, for $V = V_1$, V_2, or $V_1 \otimes_F V_2$, we know that $\dim_C(\mathrm{Hom}_F^{\mathrm{diff}}(V, E)) = \dim_F(V)$ for V_2, or $V_1 \otimes_F V_2$. By Theorem 2.4 this means $E \otimes_F V \cong E^{(m)}$ where $m = \dim_F(V)$ for V_2, or $V_1 \otimes_F V_2$. If V_1 has dimension p and V_2 has dimension q then to verify the bijection comes down to considering

$$\mathrm{Hom}_E^{\mathrm{diff}}(E^{(p)}, E) \otimes_C \mathrm{Hom}_E^{\mathrm{diff}}(E^{(q)}, E) \to \mathrm{Hom}_E^{\mathrm{diff}}(E^{(p)} \otimes_E E^{(q)}, E).$$

Let x_i, $1 \le i \le p$ and y_j, $1 \le j \le q$ be the standard bases of $E^{(p)}$ and $E^{(q)}$ and let g_i, $1 \le i \le p$ and h_j, $1 \le j \le q$ be the dual bases. These are linearly independent over C (even over E). Since $g_i \in \mathrm{Hom}_E^{\mathrm{diff}}(E^{(p)}, E)$ and $h_j \in \mathrm{Hom}_E^{\mathrm{diff}}(E^{(q)}, E)$, they form C bases of these. Since $x_i \otimes y_j$ is a basis of $E^{(p)} \otimes E^{(q)}$ and $(g_i \otimes h_j)(x_s \otimes y_t) = \delta_{is}\delta_{jt}$ is the dual basis, we have, explicitly, that the above map is bijective.

To prove the second assertion, we start with a closed, normal, coaffine subgroup H of Π which acts trivially on both W_1 and W_2. Let $G = \Pi/H$. As in the proof of Proposition 2.3, we can identify $\mathrm{Hom}_\Pi(W_i, R^{PV})$ with $F \otimes_C \mathrm{Hom}_C(W_i, C[G])$ and then $\mathrm{Hom}_C(W_i, C[G])$ with $\mathrm{Hom}_C(W_i, C)$. Then the second assertion of the proposition comes down to verifying the bijection

$$\mathrm{Hom}_C(W_1, C) \otimes_C \mathrm{Hom}_C(W_2, C) \to \mathrm{Hom}_C(W_1 \otimes_C W_2, C)$$

which is again the fact that the tensor product of duals is the dual of the tensor product for finite-dimensional vector spaces over fields.

As a further application of the category equivalence, we establish the following criterion for a differential module to be cyclic.

Proposition 2.9 *Let W be a C finite-dimensional rational Π module. Suppose there is an injective Π morphism $W \to F^{PV}$. Then the F finite-dimensional differential module $Hom_\Pi(W, F^{PV})$ is cyclic.*

Proof. Suppose $h : W \to F^{PV}$ is injective. Let $V_0 = \sum_i FD^i(h)$ be the cyclic submodule of $V = \mathrm{Hom}_\Pi(W, F^{PV})$ generated by h. If $V = V_0$ then V is cyclic. If not then V/V_0 is a non-zero finite-dimensional differential module, say of F dimension $m > 0$. Then $\mathrm{Hom}_F^{\mathrm{diff}}(V/V_0, F^{PV})$ is of C dimension m, and so in particular not zero. Any non-zero homomorphism in it gives rise by composition with $V \to V/V_0$ a differential homomorphism $f : V \to F^{PV}$ with $f \neq 0$ and $f(h) = 0$. Now

$$f \in \mathrm{Hom}_F^{\mathrm{diff}}(\mathrm{Hom}_\Pi(W, F^{PV}), F^{PV}) \cong W$$

so there is a $w \in W$ such that $\hat{w} = f$. Since $f \neq 0$, $\hat{w} \neq 0$, so $w \neq 0$. However $0 = f(h) = \hat{w}(h) = h(w)$. This in contrary to the injectivity of h. We conclude that $V = V_0$ so V is cyclic.

The converse of Proposition 2.9 is also true:

Proposition 2.10 *Suppose V is a cyclic F finite-dimensional differential module. Then there is a Π injection $Hom_F^{diff}(V, F^{PV}) \to F^{PV}$.*

Proof. Suppose that a is a cyclic generator of V. Let $W = \mathrm{Hom}_F^{\mathrm{diff}}(V, F^{PV})$ and consider $\hat{a} \in \mathrm{Hom}_F^{\mathrm{diff}}(\mathrm{Hom}_\Pi(W, F^{PV}), F^{PV}) \cong V$. Suppose there is $w \in W$, where $w : V \to F^{PV}$ is a differential homomorphism such that $\hat{a}(w) = 0$. Then $w(a) = \hat{a}(w) = 0$. Since a differentially generates V, this means that $w(v) = 0$ for all $v \in V$, so $w = 0$. Thus $\hat{a} : W \to F^{PV}$ is injective.

Both Proposition 2.9 and Proposition 2.10 have counterparts, with essentially the same proofs, where we start in the opposite category.

Proposition 2.11 *Let V be a F finite-dimensional differential module. Suppose there is an injective differential morphism $V \to F^{PV}$. Then the finite-dimensional rational Π module $Hom_F^{diff}(V, F^{PV})$ is cyclic.*

Proof. Suppose $h : V \to F^{PV}$ is injective. Let W_0 be the cyclic submodule of $W = \mathrm{Hom}_F^{\mathrm{diff}}(V, F^{PV})$ generated by h. If $W = W_0$ then W is cyclic. If not then W/W_0 is a non-zero finite-dimensional Π module, say of C dimension $m > 0$. Then $\mathrm{Hom}_\Pi(W/W_0, F^{PV})$ is of F dimension m, and so in particular not zero. Any non-zero homomorphism in it gives rise by composition with $W \to W/W_0$ a Π homomorphism $f : W \to F^{PV}$ with $f \neq 0$ and $f(h) = 0$. Now

$$f \in \mathrm{Hom}_\Pi(\mathrm{Hom}_F^{\mathrm{diff}}(V, F^{PV}), F^{PV}) \cong V$$

so there is a $v \in V$ such that $\hat{v} = f$. Since $f \neq 0$, $\hat{v} \neq 0$, so $v \neq 0$. However $0 = f(h) = \hat{v}(h) = h(v)$. This in contrary to the injectivity of h. We conclude that $W = W_0$ so W is cyclic.

and the converse to Proposition 2.11:

Proposition 2.12 *Suppose W is a cyclic finite-dimensional rational Π module. Then there is a differential injection $Hom_\Pi(W, F^{PV}) \to F^{PV}$.*

Proof. Suppose that a is a cyclic generator of W. Let $V = \mathrm{Hom}_\Pi(W, F^{PV})$ and consider $\hat{a} \in \mathrm{Hom}_\Pi(\mathrm{Hom}_F^{\mathrm{diff}}(V, F^{PV}), F^{PV}) \cong W$.

Suppose there is $v \in V$, where $v : V \to F^{PV}$ is a Π homomorphism such that $\hat{a}(v) = 0$. Then $v(a) = \hat{a}(v) = 0$. Since a generates W as a Π module, this means that $v(w) = 0$ for all $w \in W$, so $w = 0$. Thus $\hat{a} : V \to F^{PV}$ is injective.

The most important of these Propositions is Proposition 2.9. To see why, we investigate the condition that a Π module W can be embedded in F^{PV}. To this end, we start with a finite-dimensional Π module W, say of C dimension n. As we have observed before, there is a closed coaffine normal subgroup H of Π which acts trivially on W (for instance, the kernel of $\Pi \to GL(W)$) so we can regard W as a $G = \Pi/H$ module. Let W_t denote W with the trivial G action. There is a G module injection (the coaction) $W \to W_t \otimes C[G]$ by $w \mapsto \sum_i w_i \otimes f_i$ if $g(w) = \sum_i f_i(g) w_i$ for all $g \in G$. In other words, any finite-dimensional G module has a G injection into a direct sum of a finite number of copies of $C[G]$. With this observation made, we can draw the following consequence of Proposition 2.9:

Theorem 2.7 *Suppose $C \subsetneq F$. Then every F finite-dimensional differential module is cyclic, and conversely.*

Proof. By Proposition 2.9 it will suffice to prove that every C finite-dimensional rational Π module W admits an injection $W \to F^{PV}$. Choose a closed coaffine normal subgroup H of Π which acts trivially on W and let $G = \Pi/H$. A Π morphism $W \to F^{PV}$ will have image in $(F^{PV})^H$, and we know this latter is isomorphic to $F \otimes C[G]$ as a Π module. Thus we need to find a Π injection $W \to F \otimes C[G]$. The assumption that $C \subsetneq F$ implies (since C is algebraically closed) that F is infinite dimensional over C. Since Π is acting trivially on F any finite direct sum of copies of $C[G]$ appears in $F \otimes C[G]$. Since we know there is a Π module injection from W into some finite number of copies of $C[G]$, we have the desired Π injection $W \to F \otimes C[G] \cong (F^{PV})^H \subset F^{PV}$.

Suppose conversely that every differential module is cyclic. Let W be a finite-dimensional rational Π module. Then the differential module $\mathrm{Hom}_\Pi(W, F^{PV})$ is cyclic, so by Proposition 2.10 $W \cong \mathrm{Hom}_F^{\mathrm{diff}}(\mathrm{Hom}_\Pi(W, F^{PV}), F^{PV})$ injects into F^{PV} as a Π module. Choose a closed coaffine normal subgroup H of Π which acts trivially on W. Then the Π injection $W \to F^{PV}$ is actually an injection $W \to (F^{PV})^H$ and since $(F^{PV})^H \cong F[G]$ as Π modules we end up with $W \to F[G]$ an injection of Π modules. But if $C = F$ this means a G injection $W \to C[G]$. But this is clearly not the case for $W = C^{(n)}$ with trivial G action for $n > 1$. Thus if every differential module is cyclic, $C \subsetneq F$.

2.5 DIFFERENTIAL EQUATIONS WITH CONSTANT COEFFICIENTS

Theorem 2.7 shows that the differential modules over C are not all cyclic. We can still say a lot about them, using the fact that C is algebraically closed. Let V be a finite-dimensional differential C module. If we fix a basis of V we can write $V \cong (C^{(n)}, {}_AD)$ for some $A = [c_{ij}] \in M_n(C)$. Since ${}_AD(x) = x' + xA$, and $x' = 0$, ${}_AD$ is just the linear transformation $x \mapsto xA$. Thus we simplify notation and write $(C^{(n)}, A)$ for $(C^{(n)}, {}_AD)$. As C is algebraically closed, we can change basis further to assume that A is in Jordan canonical form, say with Jordan blocks $J_1, \ldots, J_m$ down the main diagonal. This means $(C^{(n)}, A)$ is differentially isomorphic to $\oplus_{i=1}^m (C^{(n_i)}, J_i)$.

We look at each summand separately, so we assume in $(C^{(n)}, A)$ that A is a single Jordan block with eigenvalue $c \in C$. That is, using the standard basis $\{e_1, \ldots, e_n\}$ for $C^{(n)}$, that $e_i A = ce_i + e_{i+1}$ for $1 \leq i \leq n-1$ and $e_n A = ce_n$.

We want to see this module in terms of differential equations. Actually we consider two equations: $D^n = 0$ and $D - \alpha$ for $\alpha \in C$. With these we associate the following (cyclic) differential modules $V(n) = Cv_1 \oplus Cv_2 \oplus \cdots \oplus Cv_n$ with derivation $D(v_i) = v_{i+1}$ for $1 \leq i \leq n-1$ and $D(v_n) = 0$, and $W(\alpha) = Cw$ with $D(w) = \alpha w$. Then we consider the modules $V(n, \alpha) = V(n) \otimes W(\alpha)$ with basis $\{y_1 = v_1 \otimes w, \ldots, y_n = v_n \otimes w\}$ and derivation $D(y_i) = D(v_i \otimes w) = v_{i+1} \otimes w + v_i \otimes \alpha w = \alpha y_i + y_{i+1}$ for $1 \leq i \leq n-1$ and $D(y_n) = D(v_n \otimes w) = v_n \otimes \alpha w = \alpha y_n$.

Thus for A the Jordan block with eigenvalue c, we have $(C^{(n)}, A) = V(n, a) = V(n) \otimes W(a)$. Let $E(n) \subset C^{PV}$ be the Picard–Vessiot extension of C for $D^n = 0$, let $E(\alpha) \subset C^{PV}$ be the Picard–Vessiot extension of C for $D - \alpha$, and let $E(n, \alpha) \subset C^{PV}$ be their compositum. The differential homomorphisms $V(n) \to C^{PV}$ (respectively $W(\alpha) \to C^{PV}$) arre given by the maps $v_1 \mapsto v$ (resppectively $w \mapsto z$) where $D^n(v) = 0$ (respectively $D(z) = \alpha z$). It follows that $E_{V(n)} = E(n)$ (respectively $E_{W(\alpha)} = E(\alpha)$) and hence that $E(n, \alpha)$ is a splitting field for $V(n, a) = (C^{n0}, A)$. Thus the compositum E of all the $E(n, \alpha)$'s splits every differential C module. By Proposition 2.6 we have that $C^{PV} = E$. Before recording this result, we examine $E(n)$ further.

Consider $E(2)$, the Picard–Vessiot extension for $D^2 = 0$. Let $W = \{y \in E(2) | D^2(y) = 0\}$. We know that W is a two-dimensional C vector space. Since $D^2(1) = 0$, we have $1 \in W$. Let $y \in W$, $y \notin C1$. Then $D(y) \neq 0$. Let $x = D(y)^{-1} y$. Then $D(x) = 1$. Also $W = C1 + Cx$. We consider the elements $v_0 = 1, v_1 = x, v_2 = (2!)^{-1}x^2, \ldots, v_n = (n!)^{-1}x^n$. For $k < 0$, let $x_k = 0$. Then $D(v_i) = v_{i_1}$ for $0 \leq i \leq n$. Then $D^i(v_i) = v_0 = 1$ and $D^{i+1}(v_i) = 0$ for all i. Suppose $\sum_{i=0}^{n} c_i v_i = 0$, $c_i \in C$. Apply D^n to the sum. Then $0 = D^n(v) = c_n v_0 = c_n$. So $0 = \sum_{i=0}^{n-1} c_1 v_1$. Apply D^{n-1} to the sum. Then $0 = c_{n-1}v_0 = c_{n-1}$. Repeating, we see that $c_0 = 0, \ldots, c_n = 0$. Thus the x_i are linearly independent over C. It follows that $Cv_0 + Cv_1 + \cdots + Cv_{n-1}$ is a full set of solutions for $D^n = 0$. Since all $v_i \in E(2)$ we have $E(2) = E(n)$ for all n. As a differential field, $E(2)$ is generated by x so $E(2) = C(x)$ and since C is algebraically closed $C(x)$ is rational functions in x. As far as the fields $E(\alpha)$ go, there will be some inclusion relations among their composites. We know, however, that each is generated for C by a single element z_α which satisfies $D(z_\alpha) = \alpha z_\alpha$. We sum up this discussion in a theorem:

Theorem 2.8 *$C^{PV} = C(x, \{z_\alpha | \alpha \in C\})$ where $D(x) = 1$ and $D(z_\alpha) = \alpha z_\alpha$. Every finite-dimensional C differential module is a sum of (cylic) modules of the form $V(n) \otimes W(\alpha)$ where $V(n) = Cv_1 \oplus \ldots Cv_n$ with derivation $D(v_1) = v_{i+1}$ for $1 \leq i \leq n-1$ and $D(v_n) = 0$, and $W(\alpha) = Cw$ where $D(w) - \alpha w$.*

CHAPTER 3

Differential Rings over Differential Fields

3.1 D FINITE ALGEBRAS AND MODULES

In Chapter 2, we studied (finite dimensional) differential modules over differential fields F (of characteristic zero and with algebraically closed fields of constants). If a monic linear homogeneous differential operator L of order n was specified, the Picard–Vessiot field extension $E_L \supseteq F$, the Picard–Vessiot ring R_L of which E_L is the quotient field, and the localized polynomial ring $S_L = F[\{x_{ij}|1 \leq i, j\}][\det^{-1}]$ of which R_L is a quotient all figured, as did the union R^{PV} of the R_L's inside the union F^{PV} of all the Picard–Vessiot extensions of F. All these differential rings S_L, R_L, and R^{PV} have the property that they are F algebras and each element f of them satisfies $\dim_F(\sum_i FD^i(f))$ is finite. (Any element f of a differential F module M is called D *finite* if it satisfies this criterion.) Note that S_L, R_L, and R^{PV} are, in general infinite-dimensional F modules. In this chapter, we will be considering infinite-dimensional (more

DOI: 10.1201/9781003452164-3

precisely, not necessarily finite dimensional) differential F modules with the property that all their elements are D finite. In addition to being F modules of this type, S_L, R_L, and R^{PV} are F algebras, and as such have modules of their own. We will be considering such modules which are also differential F modules, all of whose elements are D finite, and such that the derivation in the module is compatible with the derivation of the algebra. The algebras S_L, R_L, and R^{PV} as modules over themselves, are examples. Usually the next simplest example of a module is an ideal, but we know in the case of R_L and R^{PV} that there are no non-trivial differential ideals. As we shall see, they nonetheless have an interesting (differential-with-all-elements–D–finite) module theory.

To begin, however, we take a look at the category $\mathcal{C}_F^{\text{diff}}$ of all differential F modules and all F linear differential morphisms between them. As we have seen previously, if V and W are differential F modules (of any F dimension, including infinite), then so are $V \otimes W$ $(D(v \otimes w) = D_V(v) \otimes w + v \otimes D_W(w))$, $V \times W$ $(D((v,w)) = (D_V(v), D_W(w))$, and $\operatorname{Hom}_F(V,W)$ (if $T : V \to W$ is F linear then $D(T)(v) = D(T(v)) - T(D(v)))$. Analogous formulas apply to arbitrary direct limits of differential modules, and hence to infinite tensor products and infinite direct sums, and to arbitrary projective limits, and hence to infinite products. We conclude that $\mathcal{C}_F^{\text{diff}}$ is an $AB5$ category, an abelian category with arbitrary coproducts and products, and hence enough injectives. We can make this observation another way.

Consider more generally the category $\mathcal{C}_R^{\text{diff}}$ of differential R modules and differential R module homomorphisms for any commutative differential ring R. The *ring of differential polynomials* over R, denoted $R[X; D_R]$ is the ordinary polynomial ring over R in X with a multiplication given by the associative and distributive laws and the formula $Xa = aX + D(a)$ for $a \in R$. If M is a differential R module then M is a left $R[X; D_R]$ module under, for $r \in R$ and $m \in M$, $r \cdot m = rm$ and $Xm = D_M(m)$. Since $X(rm) = D(rm) = D(r)m + rD(m) = D(r)m + rXm$ and $(Xr)m = (rX + D(r))m = rXm + D(r)m$ we have $X(rm) = (Xr)m$; the rest of the axioms (distributive laws) for module action are easy to verify. Conversely, if M is a left $R[X; D]$ module, M is a differential R module with derivation $D_M(m) = Xm$. This applies to $R[X; D]$ itself, of course. As a differential module, $R[X; D]$ is free on the single generator $X^0 = 1$: for if M is any differential R module and $m \in M$ then $\sum a_i X^i \mapsto \sum a_i D^i(m)$ defines a (in fact unique) differential homomorphism $R[X; D] \to M$ with $1 \mapsto m$.

We will elaborate more on $R[X; D]$ as a differential R module below. Thus we can identify the category $R[X; D_R] - \text{Mod}$ of left $R[X; D_R]$ modules with $\mathcal{C}_R^{\text{diff}}$ so all the properties of the former as a category are also possessed by the latter. Thus we have the existence of infinite products and coproducts, direct and inverse limits, injectives and projectives, etc. for $\mathcal{C}_R^{\text{diff}}$ just as we have for any ring.

The previous comments apply, of course, to the case that $R = F$ is a differential field, and hence to the category $\mathcal{C}_F^{\text{diff}}$. To be able to exploit the results of the previous chapter, which usied the study of solutions of monic, linear, homogeneous differential equations with coefficients in F, we will look instead at the full subcategory $\mathcal{C}_F^{\text{findiff}}$ whose objects are the differential F modules every element of which is D finite. We call such a differential F module *completely* D finite. For example, when it is defined (i.e., when F^D is algebraically closed and F has characteristic zero) then R^{PV} is completely D finite while F^{PV} isn't. In fact R^{PV} consists of precisely all the D finite elements of F^{PV}. In general, given a differential F module V, we can consider the subset V^{fin} of all the D finite elements of V. We have defined an element $v \in V$ to be D finite provided $\dim_F(\sum FD^i v) < \infty$. Alternatively, we can say v is D finite provided it belongs to an F finite-dimensional differential submodule of V (if $v \in V_0$ and $\dim_F(V_0) < \infty$ then since $\sum FD^i v \subseteq V_0$ also $\dim_F(\sum FD^i v) < \infty$). Thus every F finite dimensional differential submodule of V consists of D finite elements. Since the sum of two F finite-dimensional differential submodules of V is again an F finite-dimensional differential submodule, it follows that the union V^{fin} of all F finite-dimensional differential submodules of V is a differential submodule, by definition a completely D finite one. And V^{fin} is also the maximal completely D finite submodule of V.

As an example, we can consider the differential module $F[X;D]$. Suppose $W \subset F[X;D]$ is a finite-dimensional differential submodule. Then not every X^i is an element of W, say $f = X^n + a_{n-1}X^{n-1} + \cdots + a_0$ is a monic element of W of maximal degree, unless $W = 0$. Then $D(f) = X^{n+1} +$ *lower degree* is a higher degree element of W. So $W = 0$ and $F[X;D]^{\text{fin}} = 0$.

If $f : V \to W$ is a differential homomorphism, the image under f of a finite-dimensional differential submodule of V is a finite-dimensional differential submodule of W. It follows that $f(V^{\text{fin}}) \subseteq W^{\text{fin}}$, i.e., that if $v \in V$ is D finite, $f(v) \in W$ is D finite.

If for $v \in V$ $\dim_F(\sum FD^i v) \leq n$ then $D^n v = \sum_0^{n-1} a_i D^i v$ for some a_i's in F, and if conversely such a relation holds then $v, Dv, \ldots, D^{n-1}$ span $\sum FD^i v$ so $\dim_F(\sum FD^i v) \leq n$. Thus we can say that $v \in V$ is D finite if and only if it is the solution of a monic linear homogeneous differential equation over F, V is completely D finite if and only if every element of V satisfies a monic linear homogeneous differential equation over F, and V^{fin} is the subset of all elements of V which satisfy a monic linear homogeneous differential equation over F. Regarding the latter characterization, a special case is the equation $Dv = 0$, namely the constants, implying that $FV^D \subseteq V^{\text{fin}}$.

There are some additional notations we will use for differential modules and for completely D finite-differential modules. As noted, a differential module over F is an $F[X;D]$ module. We abbreviate this by saying such a differential F module is an $F[D]$ module, and write $F[D]$-mod for the category $\mathcal{C}_F^{\text{diff}}$. Parallel to this notation, we say a completely D finite F differential module is an $F \cdot D$ module, and write $F \cdot D$-mod for the category $\mathcal{C}_F^{\text{findiff}}$.

The category of $F[D]$ modules is closed under direct sum. Suppose V and W are differential F modules and $(v,w) \in V \times W$. If both v and w are D finite, say $V \in V_0 \subseteq V$ and $w \in W_0 \subseteq W$ where both V_0 and W_0 are finite-dimensional differential submodule then (v,w) belongs to the finite-dimensional submodule $V_0 \times W_0$ of $V \times W$, so (v,w) is D finite. We have shown that $V^{\text{fin}} \times W^{\text{fin}} \subseteq (V \times W)^{\text{fin}}$. Conversely, if $(v,w) \in V \times W$ is D finite, so are its images $v \in W$ and $w \in W$ under the projections $V \times W \to V$ and $V \times W \to W$, since the projections are differential homomorphisms. Thus $(V \times W)^{\text{fin}} \subseteq V^{\text{fin}} \times W^{\text{fin}}$. We already have the reverse inclusion so $(V \times W)^{\text{fin}} = V^{\text{fin}} \times W^{\text{fin}}$.

The category of $F[D]$ modules is closed under tensor product. So is the category of $F \cdot D$ modules:

Proposition 3.1 *Suppose that V and W are differential modules. Then $V^{fin} \otimes_F W^{fin}$ is completely D finite, so $V^{fin} \otimes_F W^{fin} \subseteq (V \otimes W)^{fin}$.*

Proof. Suppose $x = \sum_{i=1}^{n} v_i \otimes w_i \in V \otimes W$ with $v_1, \ldots, v_n \in V^{\text{fin}}$ and $w_1, \ldots, w_n \in W^{\text{fin}}$. Let $V_0 \subseteq V$ be a finite-dimensional differential submodule with all $v_i \in V_0$ and let $W_0 \subseteq W$ be a finite-dimensional differential submodule with all $w_i \in W_0$. Then $V_0 \otimes W_0$ is a finite-dimensional differential submodule of $V \otimes W$ containing x. Hence x is D finite and $V^{\text{fin}} \otimes_F W^{\text{fin}}$ is completely D finite. In particular $V^{\text{fin}} \otimes_F W^{\text{fin}} \subseteq (V \otimes W)^{\text{fin}}$.

The situation with Hom is different, as we shall see. Let V and W be differential F modules, and consider the differential F module $\text{Hom}_F(V, W)$. We can look at its submodule of D finite elements, $\text{Hom}_F(V, W)^{\text{fin}}$. If both V and W are finite dimensional, then so is $\text{Hom}_F(V, W)$ so $\text{Hom}_F(V, W)^{\text{fin}} = \text{Hom}_F(V, W)$. In general, we have the following characterization of D finite homomorphisms of differential modules:

Proposition 3.2 *Let V and W be differential F modules and let $\phi : V \to W$ be a D finite homomorphism. Then there is a finite-dimensional differential F module Z, an element $z_0 \in Z$, and a differential homomorphism $\Phi : V \otimes Z \to W$ such that $\phi(v) = \Phi(v \otimes z_0)$. Conversely, if Y is a finite-dimensional differential F module, $y_0 \in Y$ any element, and $G : V \otimes Y \to W$ a differential homomorphism then $g : V \to W$ by $g(v) = G(v \otimes y_0)$ is a D finite homomorphism.*

Proof. Suppose $\phi : V \to W$ is an element of $\text{Hom}_F(V, W)^{\text{fin}}$, a D finite F linear homomorphism from V to W. Then by definition $Z = \sum F D^i \phi$ is an F finite-dimensional differential module consisting of F linear maps $V \to W$. Consider $\phi_0 : V \to V \otimes Z$ defined by $\phi_0(v) = v \otimes f$ and $\Phi : V \otimes_F Z \to W$ defined by $\Phi(v \otimes h) = h(v)$. Then ϕ factors as $\Phi \circ \phi_0$. Now ϕ_0 is just an F linear map, but Φ has the additional property that it is differential: for $D(v \otimes h) = D(v) \otimes h + v \otimes D(h)$ so $\Phi(D(v \otimes h)) = h(D(v)) + D(h)(v)$; since $D(h)(v) = D(h(v)) - h(D(v))$ we have $\Phi(D(v \otimes h)) = D(h(v)) = D(\Phi(v \otimes h))$. Thus the D finite-linear transformation $\phi : V \to W$ factors into a linear transformation $\phi : V \to V \otimes Z$ where Z is a finite dimensional differential module, $z_0 \in Z$, and $\phi_0(v) = v \otimes z_0$, followed by a differential homomorphism $\Phi : V \otimes Z \to W$.

Conversely, suppose we are given a finite-dimensional differential module Y, an element $y_0 \in Y$, the F linear transformation $g_0 : V \to V \otimes Y$, given by $g_0(v) = v \otimes y_0$, and a differential homomorphism $G : V \otimes Y \to W$. Then we claim that $g = G \circ g_0$ is D finite. We start by choosing a basis $y_1, \ldots, y_n$ for Y. Suppose $D(y_0) = \sum a_{ij} y_j$ for suitable $a_{ij} \in F$. Let $g_i : V \to V \otimes Y$ be given by $g_i(v) = v \otimes y_i$. We compute $D(g_i)$: $D(g_i)(v) = D(g_i(v)) - g_i(D(v)) = D(v \otimes y_i) - D(v) \otimes y_i = D(v) \otimes y_i + v \otimes D(y_i) - D(v) \otimes y_i$. Thus $D(g_i)(v) = \sum a_{ij}(v \otimes y_j)$ or $D(g_i) = \sum a_{ij} g_j$. It follows that $Fg_1 + \cdots + Fg_n$ is a (finite dimensional) differential submodule of $\mathrm{Hom}_F(V, V \otimes Y)$. Suppose $y_0 = \sum c_i y_i$. Then since $g_0 = \sum c_i g_i$ belongs to $Fg_1 + \ldots Fg_n$, which is of F dimension at most n there are $b_0, \ldots, b_{n-1}$ in F such that $D^n(g_0) - \sum_0^{n-1} b_i D^i(g_0) = 0$. Because G is differential $D^i(G \circ g_0) = G \circ D^i(g_0)$, and since G is also F linear we have $0 = G \circ (D^n(g_0) - \sum_0^{n-1} D^i(g_0)) = D^n(G \circ g_0) - \sum_0^{n-1} b_i D^i(G \circ g_0)$ which shows that $G \circ g_0$ is D finite.

The adjoint associativity isomorphism for vector spaces $\mathrm{Hom}_F(V \otimes_F Y, W) \to \mathrm{Hom}_F(V, \mathrm{Hom}_F(Y, W))$ respects differential structures if

V, Y, and W are differential modules, so we have an isomorphism which we also call adjoint associativity $\mathrm{Hom}_F^{\mathrm{diff}}(V \otimes_F Y, W) \to \mathrm{Hom}_F^{\mathrm{diff}}(V, \mathrm{Hom}_F(Y, W))$. Thus a D finite morphism $V \to W$ is determined by a finite-dimensional differential module Y, an element $y_0 \in Y$ and a differential homomorphism $V \to \mathrm{Hom}_F(Y, W)$. (If G is the differential homomorphism then the D finite homorphism $V \to W$ is given by $v \mapsto G(v)(y_0)$).

The characterization of D finite homomorphisms given by Proposition 3.2 has some useful consequences. We begin by using it to show that the composition of D finite homomorphisms is D finite.

Corollary 3.1 *Suppose that V_1, V_2, and V_3 are differential modules and $\phi : V_1 \to V_2$ and $\psi : V_2 \to V_3$ are D finite homomorphisms. Then $\psi \circ \phi : V_1 \to V_3$ is D finite.*

Proof. By Proposition 3.2 there are finite-dimensional differential module Y and Z, elements $y_0 \in Y$ and $z_0 \in Z$ and differential homomorphisms $\Phi : V_1 \otimes Z \to V_2$ and $\Psi : V_2 \otimes Y \to V_3$ such that $\phi(u) = \Phi(u \otimes z_0)$ and $\psi(v) = \Psi(v \otimes y_0)$. Define $\Upsilon : V_1 \otimes (Z \otimes Y)) \to V_3$ as $\Psi \circ (\Phi \otimes \mathrm{id}_Y)$. Note that Υ is the composition of differential maps and hence differential. Then $\Upsilon(v \otimes z_0 \otimes y_0) = \Psi(\Phi(v \otimes y_0) \otimes z_0) = \Psi(\phi(v) \otimes z_0) = \psi(\phi(v))$. Since $Z \otimes Y$ is a finite-dimensional differential module and $z \otimes y$ is an element of it we have $\psi \circ \phi$ expressed as the composition of $V_1 \to V_1 \otimes (Z \otimes Y)$ by $u \mapsto u \otimes (z_0 \otimes y_0)$ and a differential homomorphism $\Upsilon : V_1 \otimes Z \otimes Y \to V_3$. By Proposition 3.2 then the composition is D finite.

Another consequence of Proposition 3.2 is that D finite homomorphisms carry D finite elements to D finite elements.

Corollary 3.2 *Let V and W be differential F modules and let $\phi : V \to W$ be a D finite homomorphism. Then $\phi(V^{fin}) \subseteq W^{fin}$.*

Proof. By Proposition 3.2 there is a finite-dimensional differential module Y, element $z_0 \in Z$ and differential homomorphism $\Phi : V \otimes Z \to W$ such that $\phi(u) = \Phi(u \otimes y_0)$. Note that by Proposition 3.1 $V^{\text{fin}} \otimes Y$ consists of D finite elements so, since Φ is differential, $\Phi(V^{\text{fin}} \otimes Y) \subset W^{\text{fin}}$. As $\phi(V^{\text{fin}}) \subseteq \Phi(V^{\text{fin}} \otimes Y)$, this shows that $\phi(V^{\text{fin}}) \subseteq W^{\text{fin}}$.

Unlike the situation with tensor products demonstrated in Proposition 3.1, it need not be the case in general that $\text{Hom}_F(V^{\text{fin}}, W^{\text{fin}}) \subseteq \text{Hom}_F(V, W)^{\text{fin}}$. If, however, V is finite-dimensional (in which case, of course, $V = V^{\text{fin}}$), we have the following.

Corollary 3.3 *Let V be a finite-dimensional differential module. Then for any differential module W $Hom_F(V, W)^{fin} = Hom_F(V, W^{fin})$. Moreover, $V \otimes W^{fin} = (V \otimes W)^{fin}$.*

Proof. Since $W^{\text{fin}} \subseteq W$, $\text{Hom}_F(V, W^{\text{fin}}) \subseteq \text{Hom}_F(V, W)$ which implies

$\text{Hom}_F(V, W^{\text{fin}})^{\text{fin}} \subseteq \text{Hom}_F(V, W)^{\text{fin}}$. However, Corollary 3.2 implies that every D finite homomorphism $V \to W$ factors through W^{fin} so $\text{Hom}_F(V, W^{\text{fin}})^{\text{fin}} = \text{Hom}_F(V, W)^{\text{fin}}$. Since V is finite dimensional, $\text{Hom}_F(V, W^{\text{fin}}) \cong V^* \otimes W^{\text{fin}}$. Since V^* (the linear dual of V) is finite-dimensional it is completely D finite. Then by Proposition 3.1 $V^* \otimes W^{\text{fin}}$ is also completely D finite. Thus $\text{Hom}_F(V, W^{\text{fin}}) = \text{Hom}_F(V, W)^{\text{fin}}$. For the final equality, we identify the left-hand side with

$\text{Hom}_F(V^*, W^{\text{fin}})$ and right with $\text{Hom}_F(V^*, W)^{\text{fin}}$ and apply the equality just established.

The key step in the proof of Corollary 3.3 is the application of Corollary 3.2. For suppose V is a finite-dimensional differential module, W is an arbitrary differential module, and $f : V \to W$ is D finite. Once we know $f(V) \subseteq W^{\text{fin}}$ we know there is a finite dimensional differential submodule $W_0 \subseteq W$ with $f(V) \subseteq W_0$. Now every linear map $h : V \to W_0$ factors as $V \to V \otimes \text{Hom}_F(V, W_0) \to W_0$ where the first map is $v \mapsto v \otimes h$ and the second is $x \otimes g \mapsto g(x)$. As $\text{Hom}_F(V, W_0)$ is a finite-dimensional differential module, the second map is differential, we have h canonically displayed as a D finite homomorphism.

Note that the final equality of Corollary 3.3 supplies the reverse inclusion of Proposition 3.1 in the case that V is finite dimensional. As an application, we note the following (see Theorem 2.6):

Corollary 3.4 *Assume $F^D = C$ is algebraically closed of characteristic zero. Let V be a differential F module of F dimension n. Then $R^{PV} \otimes_F V$ is isomorphic to R^{PV^n} as differential F modules.*

Proof. We know that $F^{PV} \otimes V$ is isomorphic to $(F^{PV})^n$, so $(F^{PV} \otimes V)^{\text{fin}}$ is isomorphic to $((F^{PV})^n))^{\text{fin}}$. We know that $(F^{PV})^{\text{fin}} = R^{PV}$ and, by Corollary 3.3 that $(F^{PV} \otimes V)^{\text{fin}}$ is isomorphic to $(F^{PV})^{\text{fin}} \otimes V$. The result follows.

Corollary 3.4 has an implication for D finite homomorphisms with domain R^{PV}: if W is a differential F module and $h : R^{PV} \to W$ is a D finite homomorphism then there is a finite-dimensional differential module Z, an element z in Z, and a differential homomorphism $H : R^{PV} \otimes Z \to W$ so that $h(v) = H(v \otimes z)$, Since $R^{PV} \otimes_F V \cong R^{PV^n}$ H amounts to a sum of n differential homomorphisms $R^{PV} \to W$. We are going to see that these are usually all zero when $W = F$. First a lemma about D acting on R^{PV}

Lemma 3.1 *Assume $F^D = C$ is algebraically closed and characteristic zero. Then $D: R^{PV} \to R^{PV}$ is surjective. If W is a differential F module and $h: R^{PV} \to W$ is a surjective differential homomorphism then D_W : $W \to W$ is surjective.*

Proof. Let $y \in R^{PV}$ and let $L = D^n + \sum_0^{n-1} a_i D^i$ be a monic homogeneous linear differential operator over F such that $L(y) = 0$. For $L^{-1} \subset R^{PV}$ we have $y \in L^{-1}(0)$ and $\dim_C(L^{-1}(0)) = n$. Consider $L_1 = LD = D^{n+1} + \sum_1^n a_{i-1} D^i$. For $L_1^{-1}(0) \subset R^{PV}$ we have $\dim_C(L_1^{-1}(0)) = n+1$. If $z \in L_1^{-1}(0)$ we have $0 = L_1(z) = L(D(z))$ so D maps L_1^{-1} C linearly to $L^{-1}(0)$. The kernel of $D: R^{PV} \to R^{PV}$ is $\{h | D(h) = 0\}$, which is the one-dimensional space C. Thus $D: L_1^{-1}(0) \to L^{-1}(0)$ is a C linear transformation of a vector space of dimension $n+1$ to a vector space of dimension n with kernel of dimension 1, and hence surjective. We conclude there is a $z \in L_1^{-1}(0) \subset R^{PV}$ with $D(z) = y$. This proves the first assertion of the corollary, and the second is an immediate consequence.

Now suppose $F^D = C$ is algebraically closed of characteristic zero and D is not surjective on F (which happens, for example if $F = C$). Any differential homomorphism $R^{PV} \to F$ is either zero or surjective. If it's surjective, then by Lemma 3.1 D would be surjective on F, contrary to assumption. Thus $\mathrm{Hom}_F^{\mathrm{diff}}(R^{PV}, F) = 0$. But as the discussion prior to the Corollary makes clear, this also implies that $\mathrm{Hom}_F(R^{PV}, F)^{\mathrm{fin}} = 0$ also. Of course, $\mathrm{Hom}_F(R^{PV}, F)$ is not zero, and usually infinite dimensional, as it will be in the case $F = C$.

It is also instructive to consider $\mathrm{Hom}_F(F, F[X;D])^{\mathrm{diff}}$. Any D finite F linear homomorphism $g : F \to F[X;D]$ factors through a differential F linear map $G : F \otimes_F Z \to F[X;D]$ where Z is a finite-dimensional differential module. The image of G is a finite-dimensional differential submodule of $F[X;D]$ and as we saw above there are no non-zero finite-dimensional differential submodules of $F[X;D]$. Thus $\mathrm{Hom}_F(F, F[X;D])^{\mathrm{fin}} = 0$.

3.2 $F[X;D]$

We have seen that $\mathrm{Hom}_F(F, F[X;D])^{\mathrm{fin}} = 0$ and often $\mathrm{Hom}_F(R^{PV}, F)^{\mathrm{fin}} = 0$. The situation is quite different with $\mathrm{Hom}_F(F[X;D], *)^{\mathrm{fin}}$ and $\mathrm{Hom}_F(*, R^{PV})^{\mathrm{fin}}$. We begin with the former.

As an F vector space, $F[X;D]$ has the basis $\mathcal{X} = \{1, X, X^2, \ldots\}$. Thus for any differential module V, $\mathrm{Hom}_F(F[X;D], V) = V^{\mathcal{X}}$. We want to compare this to $\mathrm{Hom}_F(F[X;D], V)^{\mathrm{fin}}$. Using the adjoint associativity version, a D finite homomorphism $\phi : F[X;D] \to V$ is given by a differential homomorphism $\Phi : F[X;D] \to \mathrm{Hom}_F(Y, V)$, where V is a finite-dimensional differential module; and an element $y_0 \in Y$ used to compose Φ with evaluation at y_0: $\phi(p) = \Phi(p)(y_0)$. If Y has F dimension n, it is of the form (F^n, D_A) where A is an $n \times n$ F matrix, and y_0 is some column n tuple in F^n. Thus Y and y_0 are given by $n^2 + n$ elements of F. Since F^n has a basis of n elements, $\mathrm{Hom}_F(Y, V) = V^n$. Since $F[X;D]$ is free as a differential module on 1, differential homomorphisms $\Phi : F[X;D] \to \mathrm{Hom}_F(Y, V)$ are determined by $\Phi(1)$, and for any $T \in \mathrm{Hom}_F(Y, V)$ there is a Φ with $\Phi(1) = T$. Thus we can specify a D finite homomorphism $F[X;D] \to V$ by a tuple $(n, b_1, \ldots, b_n, a_{11}, \ldots, a_{nn}, v_1, \ldots, v_n)$, where n is the dimension of Y, $b_1, \ldots, b_n$ are elements of F giving y_0, $A = [a_{ij}]$ is the F matrix giving the differential structure $Y = (F^n, D_A)$ and $v_1, \ldots, v_n$ are elements of V giving the homomorphism $T : Y \to V$ such that $F[X;D] \to \mathrm{Hom}_F(Y,V)$ sends 1 to T. The union over all n of all these tuples is a way of counting $\mathrm{Hom}_F(F[X;D],V)^{\mathrm{fin}}$, from which we can see that it is typically much smaller than $\mathrm{Hom}_F(F[X;D],V) = V^{\mathcal{X}}$. For example, if F is countable and V is countable dimensional then this description shows that $\mathrm{Hom}_F(F[X;D],V)^{\mathrm{fin}}$ is countable but $\mathrm{Hom}_F(F[X;D],V)$ is not.

We begin with a characterization of D finite homomorphisms with domain $F[X;D]$. For this, we need a lemma, which is basically the same argument used in the proof of the converse of assertion of Proposition 3.2.

Lemma 3.2 *Let Y be a finite-dimensional differential module, V an arbitrary differential module, and $y_0 \in Y$. Then the map $g_0 : Hom_F(Y,V) \to V$ by $T \mapsto T(y_0)$ is D finite.*

Proof. Let $y_1, \ldots, y_n$ be a basis of Y and let $y_0 = \sum_i b_i y_i$, $b_i \in F$.For each i, define $g_i : \mathrm{Hom}_F(Y,V) \to V$ by $T \mapsto T(y_i)$, so $g_0 = \sum_i a_i g_i$.Then $D(g_i)(T) = D(g_i(T)) - g_i(D(T)) = D(T(y_i)) - D(T)(y_i) = D(T(y_i)) - D(T(y_i)) + T(D(y_i)) = T(D(y_i))$. If $D(y_i) = \sum_j a_{ij} y_j$ then $D(g_i)(T) = T(\sum_j a_{ij} y_j) = \sum_j a_{ij}) = \sum_j a_{ij} g_j(T)$. It follows that $Fg_1 + \ldots Fg_n$ is a (finite dimensional) differential submodule of $\mathrm{Hom}_F(Y,V)$ which contains g_0.

Proposition 3.3 *Let V be a differential F module. Let $p_0 \neq 0$ be an element of $F[X;D]$ and let $f : F[X;D]/F[X;D]p_0 \to V^{fin}$ be any F linear* *homomorphism. Then* $F[X;D] \to V$ *by* $\phi(p) = f(p + F[X;D])$ *is* D *finite and its image lies in* V^{fin}. *Conversely, every* D *finite homomorphism* $F[X;D] \to V^{fin}$ *has this form.*

Proof. The projection $F[X;D] \to F[X;D]/F[X;D]p_0$ is differential and hence D finite. Since $F[X;D]/F[X;D]p_0$ is F finite dimensional (its dimension is the degree of p_0) we can apply Corollary 3.3 to see that f is D finite and then apply Corollary 3.1 to see the composition of the projection and f, namely ϕ, is D finite. Since the image of f is contained in V^{fin}, so is the image of the composition ϕ.

Conversely, suppose $\phi : F[X;D] \to V$ is D finite with image contained in V^{fin}. Let Y be the (finite dimensional) cyclic differential submodule of $\text{Hom}_F(F[X;D], V)$ generated by ϕ and let $y_0 = \phi$. Let $\Phi : F[X;D] \to \text{Hom}_F(Y, V)$ be given by $\Phi(p)(y) = y(p)$. Then $D(\Phi(p))(y) = D(\Phi(p)(y)) - \Phi(p)(D(y)) = D(y(p)) - D(y)(p) = D(y(p)) - D(y(p)) + y(D(p)) = y(D(p)) = \Phi(D(p))(y)$. Thus $D(\Phi(p)) = \Phi(D(p))$ so Φ is a differential homomorphism, and $\Phi(p)(y_0) = y_0(p) = \phi(p)$. Thus ϕ factors as $g_0 \circ \Phi$ where g_0 is evaluation at y_0. Let $T : Y \to V$ be $\Phi(1)$ so $T(y) = y(1)$. Since $y \in Y$, $y = \sum_i a_i D^i(\phi)$ (finite sum). By assumption, $\phi(F[X;D]) \subseteq V^{\text{fin}}$. It follows that $D(\phi)(F[X;D]) \subseteq V^{\text{fin}}$, and then differentiating $D(\phi)$ that $D^2(\phi)(F[X;D]) \subseteq V^{\text{fin}}$; repeating shows that $D^i(\phi)(F[X;D]) \subseteq V^{\text{fin}}$ and hence $y(F[X;D]) \subseteq V^{\text{fin}}$. In particular, $y(1) \in V^{\text{fin}}$. Thus $T(Y) \subseteq V^{\text{fin}}$ so T is D finite by Corollary 3.3. Suppose $\sum_{i=0}^m a_i D^i(T) = 0$. Let $p_0 = \sum_{i=0}^m a_i X^i \in F[X;D]$. Since $\Phi(p) = p(D)(T)$, $\Phi(pp_0) = 0$ for all p, so Φ factors through $F[X;D] \to F[X;D]/F[X;D]p_0$.

The final assertion of Proposition 3.3 suggests asking whether, for a differential module V, a D finite homomorphism $\phi : F[X;D] \to V$ with

$\phi(F[X;D]p_0) = 0$ for some $p_0 \neq 0$ must have $\phi(F[X;D]) \subseteq V^{\text{fin}}$. Note that $F[X;D]/F[X;D]p_0$ is finite dimensional. By Corollary 3.3

$$\text{Hom}_F(F[X;D]/F[X;D]p_0, V)^{\text{fin}} = \text{Hom}_F(F[X;D]/F[X;D]p_0, V^{\text{fin}}),$$

so the answer to the question is in the affirmative.

Suppose V is a differential module and $\phi : F[X;D] \to V$ is D finite. Then ϕ factors, where one factor is a differential homomorphism $\Phi : F[X;D] \otimes Z \to V$ where Z is a finite-dimensional differential module. There is also an element $z_0 \in Z$ such that $\phi(p) = \Phi(p \otimes z_0)$. By adjoint associativity we can rewrite Φ as $\Psi : F[X;D] \to \text{Hom}_F(Z,V)$ $(\Phi(p \otimes z) = \Psi(p)(z))$. Thus $\phi(p) = \Phi(p \otimes z_0) = \Psi(p)(z_0)$. Now since $F[X;D]$ is the free differential module on 1, differential maps $F[X;D] \to \text{Hom}_F(Z,V)$ correspond to elements of $\text{Hom}_F(Z,V)$.

Proposition 3.4 *Let V and W be differential modules and suppose $\Upsilon : V \to W$ is a surjective differential homomorphism. Let $\phi : F[X;D] \to W$ be a D finite homomorphism. Then there is a D finite homomorphism $\psi : F[X;D] \to V$ such that $\phi = \Upsilon\psi$.*

Proof. Let Z be a finite-dimensional differential module and $z_0 \in Z$ such that $\phi(p) = \Phi(p)(z_0)$ for a suitable differential homomorphism $\Phi : F[X;D] \to \text{Hom}_F(Z,W)$. Since $\Upsilon : V \to W$ is surjective, so is $\text{hom}(Z,\Upsilon) : \text{Hom}_F(Z,V) \to \text{Hom}_F(Z,W)$. Because $F[X;D]$ is a free (hence projective) differential module, there is a differential homomorphism $\Psi : F[X;D] \to \text{Hom}_F(Z,V)$ such that $\text{hom}(Z,\Upsilon)\Psi = \Phi$. For $p \in F[X;D]$ then $\Phi(p) : Z \to W = \Psi(p) : Z \to V$ followed by $\Upsilon : V \to W$ or $\Phi(p)(z) = \Upsilon(\Psi(p)(z))$. Let $\psi : F[X;D] \to V$ be defined by $\psi(p) = \Psi(p)(z_0)$. Then $\Upsilon(\psi(p)) = \Upsilon(\Psi(p)(z_0)) = \Phi(p)(z_0) = \phi(p)$. So $\phi = \Upsilon\psi$.

Proposition 3.4 asserts the right exactness of $\mathrm{Hom}_F(F[X;D],*)\mathrm{fin}$

Corollary 3.5 *Let V and W be differential modules and $V \to W$ a surjective differential homomorphism. Then the induced differential homomorphism*

$$Hom_F(F[X;D],V)fin \to Hom_F(F[X;D],W)fin$$

is surjective.

Because $\mathrm{Hom}_F(F[X;D],*)^{\mathrm{fin}}$ is left exact for elementary reasons, Corollary 3.5 implies that it in an exact functor, as we now record:

Corollary 3.6 *$Hom_F(F[X;D],*)^{\mathrm{fin}}$ is an exact functor on the category of $F[D]$ modules.*

By definition $\mathrm{Hom}_F(F[X;D],V)^{\mathrm{fin}}$ is completely D finite for any differential module V. Thus Corollary 3.6 also says

Corollary 3.7 *$Hom_F(F[X;D],*)^{fin}$ is an exact functor from $F[D]$ modules to $F \cdot D$ modules. The restriction of $Hom_F(F[X;D],*)^{fin}$ to $F \cdot D$ modules is an exact functor on $F \cdot D$ modules*

3.3 R^{PV}

So that we can use the results of Chapter 2, we assume in this section that $F^D = C$ is algebraically closed of characteristic zero. We recall that if $V \subseteq W$ are finite-dimensional differential modules and $\phi : V \to R^{PV}$ is a differential homomorphism then there is a differential homomorphism $\Phi : W \to R^{PV}$ such that $\Phi(v)]\phi(v)$ for $v \in V$. Our first observation is that this result holds for any completely D finite modules.

Lemma 3.3 *Let $V \subseteq W$ be completely D finite-differential modules and $\phi : V \to R^{PV}$ a differential homomorphism. Then there is a differential homomorphism $\Phi : W \to R^{PV}$ such that $\Phi(v) = \phi(v)$ for $v \in V$.*

Proof. Consider the set $\mathcal{S}$ of pairs (W_i, Φ_i) where W_I is a differential submodule of W which contains V and $\Phi_i : W_i \to R^{PV}$ is a differential homomorphism such that $\Phi_i(v) = \phi(v)$ for $v \in V$. For two pairs say $(W_i, \Phi_i) \leq (W_j, \Phi_j)$ provided $W_i \subseteq W_j$ and $\Phi_j(w) = \Phi_i(w)$ if $w \in W_i$. Any chain $\mathcal{C} = \{(W_i, \Phi_i) | i \in \mathcal{I}\}$ in $\mathcal{S}$ has a maximal element (W_0, Φ_0) where $W_0 = \cup_i W_i$ and $\Phi_0(w) = \Phi_i(w)$ if $w \in W_i$. Thus $\mathcal{S}$ has a maximal element (W_1, Φ_1). Suppose $W_1 \subsetneq W$. Then there is a finite-dimensional differential module W_2 of W not contained in W_1. Consider the proper submodule $W_1 \cap W_2$ of W_2. The differential homomorphism $\Phi_1|(W_1 \cap W_2) : W_1 \cap W_2 \to R^{PV}$ then extends to a differential homomorphism $\Psi : W_2 \to R^{PV}$. Let $W_3 = W_1 + W_2$ and let $\Phi_3 : W_3 \to R^{PV}$ be defined by $\Phi_3(w_1 + w_2) = \Phi_1(w_1) + \Psi(w_2)$; since Φ_1 and Ψ agree on $W_1 \cap W_2$ this definition is unambigous. Then by construction $(W_1, \phi_1) < (W_3, \phi_3)$ which contradicts the maximality of (W_1, Φ_1). Therefore, $W_2 = W$ and $\Phi_1 : W \to R^{PV}$ satisfies $\Phi_1(v) = \phi(v)$ for $v \in V$.

We abstract part of the proof of Lemma 3.3 for later use:

Lemma 3.4 *Let I be a completely D finite-differential module. Suppose that for finite-dimensional differential modules $V_0 \subseteq W_0$ and differential morphisms $\phi_0 : V_0 \to I$ there is s a differential homomorphism $\Phi_0 : W_0 \to I$ such that $\Phi_0(v) = \phi_0(v)$ for $v_0 \in V_0$. Let $V \subseteq W$ be completely D finite-differential modules and $\phi : V \to I$ a differential homomorphism. Then there is a differential homomorphism $\Phi : W \to I$ such that $\Phi(v) = \phi(v)$ for $v \in V$.*

Another way to phrase Lemma 3.4 is that it asserts that a completely D finite-differential module I is injective in the category of completely D finite modules if $\mathrm{Hom}_F^{\mathrm{diff}}(\cdot, I)$ is an exact functor on the category of *finite-dimensional* differential modules. Since if V is finite-dimensional $\mathrm{Hom}_F^{\mathrm{diff}}(V, \cdot)$ commutes with direct limits, we then have that direct limits, and in particular infinite direct sums, of injective completely D finite modules are injective.

Lemma 3.3 also holds for D finite homomorphisms

Theorem 3.1 *Let $V \subseteq W$ be completely D finite-differential modules and $\phi : V \to R^{PV}$ a D finite homomorphism. Then there is a D finite homomorphism $\Phi : W \to R^{PV}$ such that $\Phi(v) = \phi(v)$ for $v \in V$.*

Proof. Let Y be a finite-dimensional differential module, $y \in Y$ and $\psi : V \otimes Y \to R^{PV}$ be a differential homomorphism such that $\phi(v) = \psi(v \otimes y)$. Consider $V \otimes Y$ as a submodule of $W \otimes Y$. By Lemma 3.3 there is a differential homomorphism $\Psi : W \otimes Y \to R^{PV}$ such that $\Psi(x) = \psi(x)$ for $x \in V \otimes Y$. In particular, $\Psi(v \otimes y) = \psi(v \otimes y)$ for $v \in V$. Let $\Phi : W \to R^{PV}$ be defined by $\Phi(w) = \Psi(w \otimes y)$. Then Φ is D finite and $\Phi(v) = \phi(v)$ for $v \in V$.

Theorem 3.1 implies that the functor $\mathrm{Hom}_F(*, R^{PV})^{\mathrm{fin}}$ is left exact on $F \cdot D$ modules; since it is trivially right exact, it is exact.

Corollary 3.8 *$Hom_F(*, R^{PV})^{fin}$ is an exact functor on the category of $F \cdot D$ modules.*

Lemma 3.3 shows that R^{PV} is an injective object in the category of $F \cdot D$ modules. We are going to use this property to further analyze its structure as a differential module.

The category $F\cdot$ contains infinite F dimensional objects, which may satisfy neither the ascending nor descending chain conditions. For example, consider $V = \sum_{i \in \mathbb{N}} V_i$ where for all i $V_i = F$ and for each $n \in \mathbb{N}$ let $V(i) = \{\sum x_i \in V | x_i = 0 \text{ for } i > n\}$ and $nV = \{\sum_i x_i | x_i = 0 \text{ for } i = n, 2n, 3n, \dots\}$. Then $V(1) \subset V(2) \subset \dots$ is an infinite properly increasing chain of submodules of V and $V \supset 2V \supset 4V \supset \dots$ is an infinite properly decreasing chain of submodules. On the other hand, finite F dimensional differential modules satisfy both the ascending and the descending chain condition, For a finite-dimensional differential module V this implies the existence of minimal proper submodules (which, since they have no proper submodules of their own are simple) and maximum proper submodules (the quotient by one of these latter can't have any submodules so it is simple) as well as slightly more refined such propers. For instance:

Lemma 3.5 *Let V be a finite-dimensional differential module and $v \in V$, $v \neq 0$. Then there is a differential homomorphism $f : V \to R^{PV}$ such that $f(v) \neq 0$.*

Proof. Let W be a submodule of V maximal with respect to the property that $v \notin W$. Consider V/W: it has the property that $v + W$ is a non-zero element and every non-zero submodule must contain $v + W$. If there is a differential morphism $g : V/W \to R^{PV}$ with $g(v + W) \neq 0$ then defining f to be the composition of g and the quotient map $V \to V/W$ satisfies the lemma, we will replace V by V/W and v with $v + W$ so we can assume that every non-zero submodule of V contains v. In particular this is true for a simple differential submodule V_0. We know that $\mathrm{Hom}_F^{\mathrm{diff}}(V_0, R^{PV})$ is non-zero (in fact has C dimension equal to the F

dimension of V_0) so that there is non-zero differential homomorphism $f_0 : V_0 \to R^{PV}$. Because V_0 is simple f_0 is a monomorphism and since $v \in V_0$ $f_0(v) \neq 0$. From exactness of $\mathrm{Hom}_F^{\mathrm{diff}}(*, R^{PV})$ on finite-dimensional modules we know that there is $f : V \to R^{PV}$ such that $f|V+0 = f_0$. Thus $f(v) \neq 0$ as desired.

We note for future reference that we showed in the course of the proof of Lemma 3.5 that if V is a simple differential module then there is a differential monomorphism $V \to R^{PV}$. Then f defines a differential bijection $\mathrm{End}_F^{\mathrm{diff}}(V) \to \mathrm{Hom}_F^{\mathrm{diff}}(V, f(V))$. Since $\mathrm{Hom}_F^{\mathrm{diff}}(V, f(V))$ is a subspace of $\mathrm{Hom}_F^{\mathrm{diff}}(V, R^{PV})$ and this latter is finite dimensional over C, so is the former and hence so is $\mathrm{End}_F^{\mathrm{diff}}(V)$. Since V is simple, this endomorphism ring is a division ring in which C is central. Over an algebraically closed field like C, the only finite-dimensional division ring with C in its center is C itself. Thus as a corollary of Lemma 3.5 we have:

Corollary 3.9 *Let V and W be simple differential modules.*

Then $\dim_C(\mathrm{Hom}_F^{diff}(V, W)) \leq 1$, with equality if and only if V and W are isomorphic.

Using Lemma 3.3 we can drop the finite-dimensional restriction in Lemma 3.5: given $v \in V$, where V is any completely D finite module, and $v \in V$, $v \neq 0$, we know there is a finite-dimensional submodule V_1 of V which contains v. We apply Lemma 3.5 to V_1 to obtain $f_1 : V_1 \to R^{PV}$ with $f_1(v) \neq 0$, then apply Lemma 3.3 to show there is an $f : V \to R^{PV}$ with $f|V_1 = f_1$ and hence $f(v) \neq 0$. Hence:

Corollary 3.10 *Let V be completely D finite-differential module and $v \in V$, $v \neq 0$. Then there is a differential homomorphism $f : V \to R^{PV}$ such that $f(v) \neq 0$.*

Corollary 3.10 shows that R^{PV} is an (injective) cogenerator of the category of $F\cdot$ modules. It's in fact a stronger assertion: to be a cogenerator only requires that if module is non-zero there is a non-zero homomorphism from the module to the cogenerator.

Recall that a non-zero submodule V_0 of a differential module V is *essential* if every non-zero submodule W of V has non-zero intersection with V_0. The *socle* of a module V (denoted $\mathrm{soc}(V)$) is the sum of all the simple submodules of V. As with any module which is the sum of simple submodules, a socle is a direct sum of some of them as we will prove in Lemma 3.7 below. The socle of a completely D finite module V is essential: for every non–zero submodule contains a non-zero finite-dimensional submodule which in turn contains a non-zero simple submodule which is contained in the socle. The socle of an infinite-dimensional completely D finite module can be infinite dimensional ($\oplus_i V_i$ with $V_i = F$ is its own socle). Or the opposite extreme: its socle could be a single simple module. We consider this possibility. One property such a module has is indecomposability.

Lemma 3.6 *Let W be a simple differential submodule of the completely D finite-differential module V. Suppose $\mathrm{soc}(V) = W$. Then V is indecomposable.*

Proof. Suppose V_1 and V_2 are submodules of V with $V = V_1 + V_2$ and $V_1 \cap V_2 = 0$. Let $p_i : V \to V_1$ be the projection. If $p_1(W) = 0$ then $W \subseteq V_2$. Thus $V_1 \cap W = 0$. Since W is an essential submodule of V, this means $V_1 = 0$. So suppose $p_1(W) \neq 0$ and $p_2(W) \neq 0$. Then the $p_i(W)$ are (non-zero) simple submodules of V_i and hence of V with $p_1(W) \cap p_2(W) = 0$.

This means $p_1(W) \oplus p_2(W) \subseteq \operatorname{soc}(V)$ which contradicts the assumption that $\operatorname{soc}(V) = W$.

Lemma 3.7 *Let W be an $F \cdot D$ module and let $\{W_\alpha | \alpha \in \mathcal{A}\}$ be a set of simple submodules of V. Then there is a subset $\mathcal{A}_0$ of $\mathcal{A}$ such that $\sum_{\mathcal{A}_0} W_\alpha$ is direct and $\sum_{\mathcal{A}_0} W_\alpha = \sum_{\mathcal{A}} W_\alpha$.*

Proof. Let $\mathcal{C} = \{\mathcal{B} \subseteq \mathcal{A} | \sum_{\mathcal{B}} W_\beta \text{ is direct}\}$. $\mathcal{C}$ is ordered by inclusion, and unions of chains in $\mathcal{C}$ belong to $\mathcal{C}$. Thus $\mathcal{C}$ has maximal elements. Let $\mathcal{A}_0$ be one of them. If $\sum_{\mathcal{A}_0} W_\alpha \neq \sum_{\mathcal{A}} W_\alpha$ then there is a W_γ such that W_γ is not contained in $\sum_{\mathcal{A}_0} W_\alpha$. Since W_γ is simple, if its intersection with $\sum_{\mathcal{A}_0} W_\alpha$ is nonzero it is contained in $\sum_{\mathcal{A}_0} W_\alpha$. Thus the intersection is zero and $\mathcal{A}_0 \cup \{\gamma\}$ is an element of $\mathcal{C}$ strictly larger than $\mathcal{A}_0$. This is impossible since this latter is maximal, so the assumption $\sum_{\mathcal{A}_0} W_\alpha \neq \sum_{\mathcal{A}} W_\alpha$ is false and the Lemma follows.

We will need to know the socle of a direct sum:

Lemma 3.8 *Let V be an $F \cdot D$ module and M_1 and M_2 submodules. Suppose $M_1 \cap M_2 = \{0\}$. Then $soc(M_1 + M_2) = soc(M_1) + soc(M_2)$.*

Proof. Note that the sum $M_1 + M_2$ is direct. Clearly $\operatorname{soc}(M_1 + M_2) \supseteq \operatorname{soc}(M_1) + \operatorname{soc}(M_2)$. To see the reverse inclusion we begin by defining $p_i : M_1 + M_2 \to M_i$ by $p_i(m_1 + m_2) = m_i$ where $m_i \in M_i$; the p_i are differential.. Suppose M is a simple submodule contained in $M_1 + M_2$. Since $M \subseteq p_1(M) + p_2(M)$ and $p_i(M) \subseteq \operatorname{soc}(M_i)$ we have $M \subseteq \operatorname{soc}(M_1) + \operatorname{soc}(M_2)$.

We will eventually see that having a simple socle is not rare, in fact is basic. First we note that every such module injects into R^{PV}.

Lemma 3.9 *Let V be a completely D finite-differential F module with simple socle. Then there is a monomorphism $V \to R^{PV}$.*

Proof. Let $V_0 = \mathrm{soc}(V)$. We know there is a monomorphism $f_0 : V_0 \to R^{PV}$. Since R^{PV} is injective, we know that there is a homomorphism $f : V \to R^{PV}$ such that $f|V_0 = f_0$. Let K be the kernel of f. Since f_0 is a monomorphism, $K \cap V_0 = 0$. Since V is an essential extension of V_0 this means that $K = 0$ so that f is the claimed monomorphism.

Thus there are isomorphic copies inside R^{PV} of all modules whose socle is a given simple module W. It is possible that W has no essential extensions (W could be injective) so that there is nothing further to learn. In general, however, there is much more structure to the set of all submodules of R^{PV} with given simple socle W, in particular to maximal ones, as we now see.

Proposition 3.5 *Let I be an injective differential module and let V_0 be a simple submodule of I. Let $\mathcal{V} = \{V \subseteq E | \ soc(V) = V_0\}$. Let E be a maximal element of $\mathcal{V}$. Then E is a direct summand of I and hence injective.*

Proof. Consider $\mathcal{G} = \{G \subset I | G$ is a submodule and $G \cap E = 0\}$. Since the union of an ascending chain in $\mathcal{G}$ belongs to $\mathcal{G}$ we can apply Zorn's Lemma to find a maximal element $G_1 \in \mathcal{G}$. Then $E + G_1$ is a direct sum: let $p : E + G_1 \to R^{PV}$ be projection on the first factor followed by inclusion into I. Because I is injective, p extends to a homomorphism $q : I \to I$ with $q|(E + G_1) = p$. In particular for $e \in E$ and $f \in G_1$ we have $q(e) = p(e) = e$ and $q(f) = p(f) = 0$. Let $E_1 = q(I)$; note that $E_1 \supseteq q(E) = E$. Now let W be any non-zero finite-dimensional submodule of E_1.

There is finite-dimensional submodule W_0 of R^{PV} such that $q(W_0) = W$: first choose a finite-dimensional submodule W_1 of I such that $W \subseteq q(W_1)$. Then consider $q_1 = q|W_1 : W_1 \to q(W_1)$. Then let $W_0 = q_1^{-1}(W) \subseteq W_1$. If $W_0 \subseteq E + G_1$ then $W = q(W_0) \subseteq q(E + G_1) \subseteq q(E) + q(G_1) = E$. If not, then W_0 is not contained in G_1 so $G_1 + W_0$ properly contains G_1. Since the latter is maximal with respect to meeting E in zero, $G_2 = (G_1 + W_0) \cap E$ is non-zero. Since $G_2 \subseteq E$, $G_2 = q(G_2)$. Since $G_2 \subseteq G_1 + W_0$, $G_2 = q(G_2) \subseteq q(G_1) + q(W_0) = q(W_0) = W$. Thus $0 \neq G_2 \subseteq E \cap W$ and in particular $E \cap W$ is non-zero, This means $V_0 \subseteq W$. If W is simple, this means $W = V_0$, so $\mathrm{soc}(E_1) = W$. Since $E_1 \supseteq E$ and E is maximal with respect to submodules of I having socle W, $E_1 = E$. So $q : I \to I$ has image E; since q is the identity on E this means that q is a projection onto E as a direct summand of I. This proves the main assertion of the proposition.

The proof of Proposition 3.5 is based on a proof of the corresponding result for modules over commutative rings posted online by Lee Lady.

Theorem 3.2 *Let W be a simple submodule of R^{PV}. Then there is an injective submodule $E(W) \subseteq R^{PV}$ which contains W as its socle. Any injective module whose socle is isomorphic to W is isomorphic to $E(W)$.*

Proof. We know that R^{PV} is injective so we can apply Proposition 3.5 with $I = R^{PV}$. Zorn's lemma applied to the set $\mathcal{V}$ shows that $\mathcal{V}$ has maximal elements, and Proposition 3.5 shows that these are injective direct summands of R^{PV}. Let E_1 be one such. Suppose that E_0 is an injective module with socle $W_0 \cong W$. Let $\phi_0 : W_0 \to W$ be an isomorphism. Consider $W_0 \to E_1$ given by ϕ_0 followed by inclusion. Since E_1

is injective, this map extends from the submodule W_0 to a homomorphism $\Phi : E_0 \to E_1$. Because Φ restricted to the socle W_0 of E_0 is a monomorphism, Φ is a monomorphism whose image $\Phi(E_0)$ is an injective submodule of E_1 and hence a direct summand, say $E_1 = \Phi(E_0) \oplus G$. This is a direct sum decomposition of E_1, But E_1 has simple socle W, so E_1 is indecomposable by Lemma 3.6. Since $\Phi(E_1) \neq 0$, we have $G = 0$ and hence $\Phi(E_0) = E_1$ we already knew Φ was a monomorphism; since it is onto it is an isomorphism.

We will continue to use the notation of Theorem 3.2: $E(V)$ is a maximal submodule of R^{PV} with respect to having the simple submodule V of R^{PV} as its socle. As we will see, those conditions do not specify it uniquely so it is *a* rather than *the* maximal such submodule. We call $E(V)$ an *injective hull* of V.

The notation $E(V)$ is applied to any simple submodule of R^{PV}. Thus we have $E(F)$ for the submodule F. This is a very important such module.

Theorem 3.3 *Let V be a simple submodule of R^{PV}. Then $V \otimes_F E(F)$ is an injective differential module with socle $V \otimes_F F$ and $VE(F)$ (multiplication in R^{PV}) is (an) $E(V)$.*

Proof. Let W be a D finite-differential F module. We consider $\mathrm{Hom}_F^{\mathrm{diff}}(W, V \otimes_F E(F))$. We use the isomorphism $V \otimes_F E(F) \cong \mathrm{Hom}_F(V^*, E(F))$ and adjoint associativity to write $\mathrm{Hom}_F^{\mathrm{diff}}(W, V \otimes_F E(F))$ as $\mathrm{Hom}_F^{\mathrm{diff}}(W \otimes_F V^*, E(F))$. As a functor of W, this is the composition of the exact functors $* \otimes V^*$ and $\mathrm{Hom}_F^{\mathrm{diff}}(*, E(F))$ and hence an exact functor. This tells that $V \otimes_F E(F)$ is injective. Now suppose W is simple and that $W \to V \otimes E(F)$ is not zero. Then the corresponding map $W \otimes V^* \to E(F)$ is not zero. We will explain below in Lemma 3.10 why $W \otimes V^*$ is a sum of simple modules, say $W \otimes V^* = W_1 \oplus W_2 \oplus \cdots \oplus W_n$ with W_i simple. The homomorphisms $W_i \to E(F)$ are either 0 or injective, and if the latter image of W_i is in the socle F of $E(F)$, which means $W_i \cong F$. Thus if there is a non-zero homomorphism $W \otimes V^* \to E(F)$ then F appears as a summand of $W \otimes V^*$. Thinking of $W \otimes V^*$ as $\mathrm{Hom}_F(V, W)$ we see that having F as a summand of $W \otimes V^*$ means that $\mathrm{Hom}_F(V, W)^D = \mathrm{Hom}_F^{\mathrm{diff}}(V, W)$ is non-zero which of course means W is isomorphic to V. Thus we need only consider $\mathrm{Hom}_F^{\mathrm{diff}}(V, V \otimes_F E(F)) = \mathrm{Hom}_F^{\mathrm{diff}}(\mathrm{Hom}_F(V, V), E(F))$. As noted, the differential maps $\mathrm{Hom}_F(V, V) \to E(F)$ all factor through the F summands in $\mathrm{Hom}(V, V)$ (as a differential module), of which there is only one. Thus $\mathrm{Hom}_F^{\mathrm{diff}}(\mathrm{Hom}_F(V, V), E(F)) = \mathrm{Hom}_F^{\mathrm{diff}}(F, E(F))$ and any such homomorphism $F \to E(F)$, indeed any differential map $h : F \to R^{PV}$ has $h(F)^D \subseteq (R^{PV})^D = C$ so $h(F) = Fh(C) \subseteq FC = F$. Since $\mathrm{Hom}_F^{\mathrm{diff}}(F, F) = C$, this means $\dim_C(\mathrm{Hom}_F^{\mathrm{diff}}(V, V \otimes_F E(F))) = 1$. There is a differential monomorphism $g : V \to V \otimes E(F)$ by $v \mapsto v \otimes 1$. Thus $Cg = \mathrm{Hom}_F^{\mathrm{diff}}(V, V \otimes_F E(F))$ This then accounts for all differential homomorpisms $V \to V \otimes E(F)$ and they all have image $V \otimes F$. Thus $V \otimes F$ is the only simple submodule of $V \otimes E(F)$ and hence its socle. Now consider the differential homomorphism $V \otimes E(F) \to R^{PV}$ by $v \otimes e \mapsto ve$. This maps $V \otimes F$ to V injectively. Since $V \otimes F$ is the socle of $V \otimes E(F)$ this means the morphism is a monomomorphism. The image is an injective submodule of R^{PV} and its socle is V. Thus the image is (an) $E(V)$.

The argument used in the proof of Theorem 3.3 to show that $V \otimes_F E(F)$ is injective also shows that $V \otimes_F I$ is injective for any finite-dimensional differential module V (not just simple ones) and any injective differential module I (not just the special case $I = E(F)$). Then since $V^* \otimes_F I \cong \text{Hom}_F(V, I)$ we know the latter is also injective. Further, if W is any completely D finite-differential module, it is the direct limit of finite-dimensional differential submodules W_i, so $W \otimes_F I$ is the direct limit of its injective submodules $W_i \otimes_F I$. We know that for $F \cdot D$ modules that direct limits of injectives are injective. So as a corollary of the proof of Theorem 3.3 we have:

Corollary 3.11 *Let W and I be $F \cdot D$ modules with I injective. Then $W \otimes_F I$ is injective. If W is finite dimensional, then $Hom_F(W, I)$ is injective.*

It remains to prove the fact, used in the proof of Theorem 3.3 that the tensor product of two simple differential modules is semisimple. This is a well-known fact for modules W_1 and W_2 over an algebraic group G: if the W_i are (semi)simple then the unipotent radical U of G acts trivially on them so that they are modules for the reductive group G/U. Then $V_1 \otimes V_2$ is also a G/U module, and since G/U is reductive all its modules are semisimple. We can use the duality given by $\text{Hom}_F^{\text{diff}}(*, R^{PV})$ and $\text{Hom}_\Pi(*, R^{PV})$ to transfer this fact to differential modules. We use that the dualities carry tensor products to tensor products and that the duals of simple modules are simple; we have shown the former, and the latter is a straightforward observation. Thus:

Lemma 3.10 *Let V_1 and V_2 be (semi)simple differential modules. Then $V_1 \otimes_F V_2$ is semisimple.*

Proof. Let $W_i = \mathrm{Hom}_F^{\mathrm{diff}}(V_i, R^{PV})$. Since V_i is a simple differential module, W_i is a simple Π module. There is a closed normal subgroup H of Π such that H acts trivially on W_1 and W_2 and $G = \Pi/H$ is affine. Let U be the unipotent radical of G. Consider W_i^U: As U is normal, this is a G submodule of W_i hence either W_i or 0. Since U is unipotent, the U module V_i^U is non-zero. Thus $W_i = W_i^U$ and W_i is a module for the group G/U. Hence $W_1 \otimes_C W_2$ is also a module for G/U, and as G/U is reductive all its modules are semisimple. All G/U modules are Π modules under $\Pi \to G \to G/U$ so $W_1 \otimes W_2$ is a direct sum of simple Π modules. Applying $\mathrm{Hom}_\Pi(*, R^{PV})$ we thus conclude that $V_1 \otimes_F V_2$ is also a direct sum of simple differential modules. The parenthetical "semi" follows from the observation that a tensor product of semisimple modules is a direct sum of tensor products of simple modules.

As Theorem 3.3 shows, once we have specified a choice for $E(F)$ then for every other simple submodule V of R^{PV}, we can take $VE(F)$ for $E(V)$, thus specifying these also.

For the next part of the investigation, we need some basic facts about injective modules. We have already seen, in Lemma 3.4, that the injectivity of a module can be tested on finite modules. This immediately implies that (possibly infinite) direct sums and all direct limits of injective modules are injective. We have also used the fact that the inclusion (hence any monomorphism) of an injective module I into a module V has a left inverse (the identity $I \to I$ extends to $V \to I$ which is a left inverse to the identity). We also have the folllowing:

We have seen that for every simple submodule V of R^{PV}, we have an injective submodule $E(V) \subset R^{PV}$. In a sense to be made clear in the following results, these injective submodules cover R^{PV}.

Proposition 3.6 *$R^{PV} = \sum E(V)$ where the sum ranges over all simple submodules V of R^{PV}.*

Proof. We will establish that $E = \sum E(V)$ is injective. Let $\{V_\alpha | \alpha \in \mathcal{A}\}$ be the set of simple submodules of R^{PV}. By Lemma 3.7 we know that there is $\mathcal{B} \subseteq \mathcal{A}$ such that $\sum_{\alpha \in \mathcal{A}} V_\alpha = \sum_{\beta \in \mathcal{B}} V_\beta$, and the second sum is direct. Consider $\Phi : \oplus_{\mathcal{B}} E(V_\beta) \to \sum_{\mathcal{B}} V_\beta E(F)$ obtained from the isomorphisms $E(V_\beta) \to V_\beta E(F)$ of Theorem 3.3. By Lemma 3.8 we have $\mathrm{soc}(\oplus_{\mathcal{B}} E(V_\beta)) = \oplus_{\mathcal{B}} V_\beta$. Since Φ is a monomorphism on the socle of the domain, it is a monomorphism, and hence an isomorphism. Thus $E = \sum_{\mathcal{B}} V_\beta E(F)$ is a direct sum and an injective module, hence a direct summand of R^{PV}, say $E \oplus E_1 = R^{PV}$ for some submodule $E_1 \subset R^{PV}$. Then by Lemma 3.8 $\mathrm{soc}(R^{PV}) = \mathrm{soc}(E) + \mathrm{soc}(E_1)$ But $\mathrm{soc}(E) = \sum_{\mathcal{B}} V_\beta = \sum_{\mathcal{B}} V_\beta = \sum_{\mathcal{A}} V_\alpha = \mathrm{soc}(R^{PV})$ so $\mathrm{soc}(E_1) = 0$ so $E_1 = 0$ and $E = R^{PV}$. Since $\mathcal{B} \subset \mathcal{A}$, $E \subset \sum_{\mathcal{A}} E(V_\alpha)$ which proves the proposition.

The socle of R^{PV} is the sum of all the simple submodules of R^{PV}. This sum may be written as a direct sum of some of the simple submodules. Just as in the proof of Proposition 3.6 multiplying this direct sum by $E(F)$ means we can write the conclusion of the proposition as a direct sum statement:

Corollary 3.12 *There is a set of simple submodules $\{V_\gamma | \gamma \in \Gamma\}$ of R^{PV} such that $R^{PV} = \oplus_\gamma E(V_\gamma)$.*

In the notation of Corollary 3.12, we have $\mathrm{soc}(R^{PV}) = \oplus_\gamma V_\gamma$. If V is a simple module, we know there is a non-zero morphism $f : V \to R^{PV}$, then $f(V) \subset \oplus_\gamma V_\gamma$, so at least one of the projections $f(V) \to$

V_γ must be non-zero, and hence an isomorphism. Thus V is isomorphic to some V_γ. Let $\Gamma_0 = \{\gamma \in \Gamma | V_\gamma \cong V\}$. Then $\text{Hom}_F^{\text{diff}}(V, R^{PV}) = \text{Hom}_F^{\text{diff}}(V, \oplus_{\gamma \in \Gamma_0} V_\gamma)$. We know that $\dim_C(\text{Hom}_F^{\text{diff}}(V, R^{PV})) = \dim_F(V)$, while $\text{Hom}_F^{\text{diff}}(V, \oplus_{\gamma\in\Gamma_0} V_\gamma) = \oplus_{\gamma\in\Gamma_0}\text{Hom}_F^{\text{diff}}(V, V_\gamma)$.

Since $\dim_C(\text{Hom}_F^{\text{diff}}(V, V_\gamma)) = 1$ we conclude that $|\Gamma_0| = \dim_F(V)$. Thus the socle of R^{PV} consists of the direct sum over all isomorphism classes of simple modules of finite sums of isomorphic simple modules, each one occurring as many times as its dimension.

The description of R^{PV} as a sum of injective hulls of simple modules easily extends to any injective $F \cdot D$ module:

Theorem 3.4 *Let I be an injective $F \cdot D$ module. Let V_i, $i \in \mathcal{I}$, be simple submodules of I such that $soc(I) = \oplus_i V_i$. Then $I \cong \oplus_i E(V_i)$. If V is any simple $F \cdot D$ module, the cardinality of $\{i \in \mathcal{I} | V_i \cong V\}$ (if finite) is $dim_C(Hom_F^{diff}(V, I))$.*

Proof. The existence of the set $\mathcal{I}$ is guaranteed by Lemma 3.7. Then the same reasoning as Proposition 3.6 and Corollary 3.12 with I replacing R^{PV} provides the first assertion, and the second follows from the first.

If I, like R^{PV} is such that for all simple V the cardinality of $\{i \in \mathcal{I} | V_i \cong V\}$ is finite, then

$$I \cong \oplus_V E(V)^{(a_V)}$$

where V ranges over the isomorphism classes of simple $F \cdot D$ modules and $a_V = \dim_C(\text{Hom}_F^{\text{diff}}(V, I))$.

Lemma 3.6 shows that $E(V)$ for V simple is indecomposable, and implies that these are the only indecomposable injective modules. Thus the direct sum decomposition of Theorem 3.4 is a decomposition of I into indecomposable injectives.

3.4 INJECTIVE RESOLUTIONS

It is a consequence of Corollary 3.11 that if W is any $F \cdot d$ module then we can construct an injective resolution $I_0 \to I_1 \to \dots$ of W: that is, a long exact sequence of injective $F \cdot D$ modules I_i, $i = 0, 1, 2, \dots$ such that $\mathrm{Ker}(I_0 \to I_1) \cong W$. To do so, we first consider how to construct a resolution for F. First note that for any $F \cdot D$ module W we have an injection $W \to W \otimes_F E(F)$ by $w \mapsto w \otimes 1$. Now take $E_0 = E(F)$ and let $E_1 = E_0/F \otimes_F E(F)$ and let $f_0 : E_0 \to E_1$ be the quotient map followed by the injection, namely $e \mapsto e + V \otimes 1$. Then iterate: $E_2 = (E_1/f_0(E_0)) \otimes E(F)$ with $f_1 : E_1 \to E_2$ given by $e \mapsto (e + f_0(E_0)) \otimes 1$, etc. This defines a long exact sequence of injective modules with $\mathrm{Ker}(E_0 \to E_1) \cong F$. Thus E_i, $i = 0, 1, 2, \dots$ with transitions maps $f_i : E_i \to E_{i+1}$ is an injective resolution of F. To write some later formulas uniformly we adopt the notational convention that $E_i = 0$ if $i < 0$.We reserve this notation (E_i, f_i) for this particular injective resolution of F.

It follows that $W \otimes_F E_i$ $i = 0, 1, 2, \dots$ is an injective resolution of $W$with transition maps $1_W \otimes f_i$ for any $F \cdot D$ module W. If $\mathcal{F}$ is a functor on $F \cdot D$ modules we can use these resolutions to compute the derived functors of $\mathcal{F}$: $R^i\mathcal{F}(W) = \mathrm{Ker}(\mathcal{F}(1_W \otimes f_i))/\mathrm{Image}(\mathcal{F}(1_W \otimes f_{i-1}))$. It is a general fact from categorical homological algebra that the computation of the derived functors of a left exact (monomorphism preserving) functor do not depend on the choice of injective resolution. For example, if I is injective, $I_0 = I$ and $I_i = 0$ for $i > 0$ is an injective resolution of I. Thus $R^i\mathcal{F}(I) = \mathcal{F}(I)$ and $R^i\mathcal{F}(I) = 0$ for $i > 0$.

It is also standard categorical homological algebra that if $0 \to V \to X \to Z \to 0$ is a short exact sequence of $F \cdot D$ modules then there is a long exact sequence of derived functors

$$\ldots R^i\mathcal{F}(V) \to R^i\mathcal{F}(X) \to R^i\mathcal{F}(W) \to R^{i+1}\mathcal{F}(V) \to \ldots$$

For example, let V be an $F \cdot D$ module and let $\mathcal{F} = \mathrm{Hom}_F^{\mathrm{diff}}(V, \cdot)$. In this case it is customary to write $R^i\mathcal{F}(W)$ as $\mathrm{Ext}_F^i(V, W)^{\mathrm{diff}}$. If I_i, f_i is an injective resolution of W then $\mathrm{Ker}(\mathrm{Hom}_F^{\mathrm{diff}}(V, I_0) \to \mathrm{Hom}_F^{\mathrm{diff}}(V, I_1)) = \mathrm{Hom}_F^{\mathrm{diff}}(V, \mathrm{Ker}(f_0)) = \mathrm{Hom}_F^{\mathrm{diff}}(V, W)$. Thus $\mathrm{Ext}_F^0(V, W)^{\mathrm{diff}} = \mathrm{Hom}_F^{\mathrm{diff}}$ (V, W). Moreover, if I is injective then using the resolution $I_0 = I$, $I_i = 0$ for $i > 0$, we have $\mathrm{Ext}_F^i(V, I)^{\mathrm{diff}} = 0$ for $i > 0$.

Let W be a $F \cdot D$ and E an injective module which contains it (like $W \otimes_F E(F)$). Then the short exact sequence $0 \to W \to E \to E/W \to 0$ yields the long exact sequence which begins

$$0 \to \mathrm{Hom}_F^{\mathrm{diff}}(V, W) \to \mathrm{Hom}_F^{\mathrm{diff}}(V, E)$$

$$\to \mathrm{Hom}_F^{\mathrm{diff}}(V, E/W) \to \mathrm{Ext}_F^1(V, W)^{\mathrm{diff}} \to 0$$

This says that $\mathrm{Hom}_F^{\mathrm{diff}}(V, E/W)$ maps surjectively onto $\mathrm{Ext}_F^1(V, W)^{\mathrm{diff}}$ with the kernel being those differential maps $V \to E/W$ which come from maps $V \to E$.

To better understand what the above description of $\mathrm{Ext}_F^1(V, W)^{\mathrm{diff}}$. we consider the homomorphisms more carefully. Let $f : V \to E/W$ be a differential homomorphism. Let $X \subset E \times V = \{(e, v) | e + W = f(v)\}$. Note that X is a differential submodule of $E \times V$. (And X is the fibre product $E \times_{E/V} V$.) This construction of X is based on the construction that works in general abelian categories. In the context of modules, we can more simply say that $X = \pi^{-1}(f(V))$ where $\pi : E \to E/W$ is the projection. Let $q : X \to V$ be the projection on the second factor, which is a differential homomorphism. Since $E \to E/W$ is onto, for any $v \in V$ there is an $e \in E$ such that $e + W = f(v)$. This shows that q is onto. We are regarding $W \to E$ as the inclusion. Since $e + W = W$ for $e \in W$ the pair $w, 0)$ lies in X. Thus the map $p : E \times V$ by injection on the first factor actually has image in X. The kernel of q is the set of pairs $(e, 0) \in X$. That is $e + W = f(0) = 0$ which means $e \in W$ so the kernel is $p(W)$. Thus from f we have the exact sequence

$$0 \to W \to X \to V \to 0.$$

This gives a correspondence between homomorphisms and short exact sequences starting with W and ending with V, and thus a description of $\text{Ext}^1_F(V, W)^{\text{diff}}$ in terms of exact sequences. Of course, the homomorphisms are only defined up to those which factor through $E \to E/W$. Conversely, suppose we have an exact sequence $0 \to W \to Y \to V \to 0$.

Consider the diagram

$$\begin{array}{ccccccccc} 0 & \longrightarrow & W & \longrightarrow & Y & \longrightarrow & V & \longrightarrow & 0 \\ & & \| & & & & & & \\ 0 & \longrightarrow & W & \longrightarrow & E & \longrightarrow & E/W & \longrightarrow & 0 \end{array}$$

Because E is injective and we have $W \to E$ the monomorphism $W \to Y$ extends to a homomorphism $h : Y \to E$ which, by construction, agrees with the identity on W. Thus composing h with $E \to E/W$ vanishes on

W and hence induces a map $g : V \to E/W$ such that the diagram

$$\begin{array}{ccccccccc} 0 & \longrightarrow & W & \longrightarrow & Y & \longrightarrow & V & \longrightarrow & 0 \\ & & \| & & \downarrow h & & \downarrow g & & \\ 0 & \longrightarrow & W & \longrightarrow & E & \longrightarrow & E/W & \longrightarrow & 0 \end{array}$$

commutes. The constructions $f \mapsto X$ and $Y \mapsto g$ are not exactly inverse to each other; the choice of h in the second construction is not unique. If h_1 and h_2 are two different choices, then $\phi = h_1 - h_s$ vanishes on W and hence factors through $Y \to V$. Thus $\phi : V \to E$ and h_1 and h_2 differ by an element of $\mathrm{Hom}_F^{\mathrm{diff}}(V, E)$. Thus the exact sequence $0 \to W \to Y \to V \to 0$ determines a homomorphism $V \to E/W$ modulo homomorphisms $V \to E$. In other words, the exact sequence determines an element of $\mathrm{Ext}_F^1(V, W)^{\mathrm{diff}}$. We will explore this more thoroughly below.

For an example, suppose $W = F$, $E = E(F)$, and V is a simple differential module not isomorphic to F. If $f : V \to E(F)$ is differential and non-zero, $f(V)$, which is isomorphic to V, is in the socle F of $E(F)$, which is a contradiction. Thus $\mathrm{Hom}_F^{\mathrm{diff}}(V, E(F)) = 0$ so $\mathrm{Ext}_F^1(V, F) \cong \mathrm{Hom}_F^{\mathrm{diff}}(V, E(F)/F)$. We will see in the next section what it means to have a non-zero differential homomorphism $V \to E(F)/F$ for V simple.

To return to our long exact sequence in general, we have for $i \geq 1$ that

$$\cdots \to \mathrm{Ext}_F^i(V, E)^{\mathrm{diff}} \to \mathrm{Ext}_F^i(V, E/W)^{\mathrm{diff}}$$

$$\to \mathrm{Ext}_F^{i+1}(V, W)^{\mathrm{diff}} \to \mathrm{Ext}_F^{i+1}(V, E)^{\mathrm{diff}} \to \ldots;$$

since both $\mathrm{Ext}_F^i(V, E)^{\mathrm{diff}} = 0$ and $\mathrm{Ext}_F^{i+1}(V, E)^{\mathrm{diff}} = 0$ the above exact sequence implies that $\mathrm{Ext}_F^{i+1}(V, W)^{\mathrm{diff}} \cong \mathrm{Ext}_F^i(V, E/W)^{\mathrm{diff}}$ (this is called dimension shifting).

We look more carefully at the construction of $\mathrm{Ext}^i_F(V,W)^{\mathrm{diff}}$ when V is finite dimensional. Let E_i, f_i, be an injective resolution of F, so that $E_i = W \otimes_F I_i$, $g_i = 1 \otimes f_i$ is an injective resolution of W. Then for $i > 0$,

$$\mathrm{Ext}^i_F(V,W)^{\mathrm{diff}} = \mathrm{Ker}(\mathrm{Hom}^{\mathrm{diff}}_F(V,g_i))/g_{i-1}(\mathrm{Hom}^{\mathrm{diff}}_F(V,E_{i-1})).$$

For any differential module X, we have $\mathrm{Hom}^{\mathrm{diff}}_F(V,X) = \mathrm{Hom}_F(V,X)^D$, which in turn equals $\mathrm{Hom}^{\mathrm{diff}}_F(F,\mathrm{Hom}_F(V,X))$. This applies in particular to the case $X = E_i$.. Since $\mathrm{Hom}_F(V,\cdot)$ is an exact functor, the sequence $\mathrm{Hom}_F(V,E_i)$ is exact, except of course for $i=0$, where the kernel is $\mathrm{Hom}_F(V,W)$. Since we are assuming V is finite, $\mathrm{Hom}_F(V,E_i) = V^* \otimes_F E_i$ is an injective $F \cdot D$ module. So $\mathrm{Hom}_F(V,E_i)$ is an injective resolution of $\mathrm{Hom}_F(V,W)$. So if we apply $\mathrm{Hom}^{\mathrm{diff}}_F(F,\cdot)$ and take homology of the sequence $\mathrm{Hom}^{\mathrm{diff}}_F(F,\mathrm{Hom}_F(V,E_i))$, we get $\mathrm{Ext}^i_F(F,\mathrm{Hom}_F(V,W))^{\mathrm{diff}}$. On the other hand, since $\mathrm{Hom}^{\mathrm{diff}}_F(F,\mathrm{Hom}_F(V,E_i)) = \mathrm{Hom}^{\mathrm{diff}}_F(V,E_i)$ the homology is also $\mathrm{Ext}^i_F(V,W)^{\mathrm{diff}}$. Thus

$$\mathrm{Ext}^i_F(F,\mathrm{Hom}_F(V,W))^{\mathrm{diff}} = \mathrm{Ext}^i_F(V,W)^{\mathrm{diff}}.$$

In the case $i=1$, we can interpret the elements on both sides of the above equation as exact sequences. On the left, we have sequences

$$0 \to \mathrm{Hom}_F(V,W) \to X \to F \to 0$$

and on the right sequences

$$0 \to W \to Y \to V \to 0.$$

We can apply the exact functor $\mathrm{Hom}_F(V,\cdot)$ to the second sequence to get

$$0 \to \mathrm{Hom}_F(V,W) \to \mathrm{Hom}_F(V,Y) \to \mathrm{Hom}_F(V,V) \to 0.$$

Now $F = F \cdot 1_V$ is a differential submodule of $\mathrm{Hom}_F(V,V)$; let its inverse image in $\mathrm{Hom}_F(V,Y)$ be Z. Then we have an exact sequence

$$0 \to \mathrm{Hom}_F(V,W) \to Z \to F \to 0.$$

which gives an exact sequence of the first type.

Conversely, suppose we start we an exact sequence of the first type. We can view $\mathrm{Hom}_F(V,W)$ as $V^* \otimes W$. First we apply the exact functor of linear duals to get

$$0 \to F^* \to X^* \to W^* \otimes V \to 0.$$

then tensor with V^* to get the exact sequence

$$0 \to F^* \otimes V^* \to X^* \otimes V^* \to W^* \otimes V \otimes V^* \to 0.$$

Let x_i be a basis of V and f_i the dual basis. Then under $V^* \otimes V \cong \mathrm{Hom}_F(V,V)$ $t = \sum f_i \otimes x_i$ corresponds to the identity. Then $D(t) = 0$ so $W^* \to W^* \otimes V^* \otimes V$ by $x \mapsto x \otimes t$ is a differential inclusion. Let P be the inverse image of $W^* \otimes_F Ft$ in X^*. Then (with W^* instead of $W^* \otimes_F Ft$) we have an exact sequence

$$0 \to V^* \to P \to W^* \to 0$$

whose linear dual

$$0 \to W \to P^* \to V \to 0$$

is a sequence of the second type.

We close this section with some comments on constructing injective resolutions. If W is any completely D finite-differential module, we let $E(W)$ denote an injective module which contains an isomorphic copy of

W as an essential submodule. Despite the notation, $E(W)$ is not determined by W, but if we fix a monomorphism $W \to E(W)$ then $E(W)$ is determined up to isomorphism. Note we have already fixed $E(V)$ for V a simple submodule of R^{PV}. Then we can construct an injective resolution E^i of W by letting $E^0 = E(W)$, $E^! = E(E^0/W)$, and so forth. This construction has the advantage that if E^i is injective for any i then $E^j = 0$ for $j > i$. Or let I be any injective module which contains F as a submodule, such as $E(F)$ or R^{PV}. Then for any $F \cdot D$ module W we have a differential homomorphism $W \to W \otimes_F I$ by $w \mapsto w \otimes 1$. And we know that $W \otimes_F I$ is injective. Then we can construct a resolution as follows: $E^0 = W \otimes I$. $E^1 = (E^0/(W \otimes 1) \otimes I$, and so forth. Note that these resolutions are functorial in W. In the case $I = R^{PV}$, we have the further observation that each E^i is (non-canonical) a direct sum of copies of R^{PV}:

Lemma 3.11 *Let W be a completely D finite-differential F module. Then $W \otimes_F R^{PV}$ is F differentially isomorphic to a direct sum of copies of R^{PV}.*

Proof. When V is finite dimensional it was proven in Theorem 2.6 that $V \otimes R^{PV}$ has a basis of constants, which means it is isomorphic to $R^{PV} \oplus \cdots \oplus R^{PV}$ ($\dim_F(V)$ summands). Another way to phrase "basis of constants" is to say $R^{PV} \otimes_F V \cong R^{PV} \otimes_C (R^{PV} \otimes_F V)^D$. The map $R^{PV} \otimes_C (R^{PV} \otimes_F V)^D \to R^{PV} \otimes_F V$ by $a \otimes_C h \mapsto ah$ is compatible with direct limits in V. Since it is an isomorphism when V is finite dimensional, it is an isomophism for all V.

Because of Lemma 3.11, if we construct an injective resolution using $(\cdot) \to (\cdot) \otimes R^{PV}$ then all the modules in the resolution are direct sums of copies of R^{PV}. Suppose we are using such a resolution to compute

$\text{Ext}^i(F,W)^{\text{diff}}$. Applying $\text{Hom}_F^{\text{diff}}(F,\cdot)$ to these injectives is like applying $(\cdot)^D$; thus if E^i is a direct sum of α copies of R^{PV} then $(E^i)^D$ will be a direct sum of α copies of C.

3.5 EXTENSIONS

For much of the initial material in this section, we will phrase the discussion in terms of differential modules over a differential ring R, of which $R = F$ is then a special case. We assume that V and W are differential R modules. An *extension* of W by V is a differential module E, an injective differential homomorphism $p: V \to E$, and a surjective differential homomorphism $q: E \to W$ such that $\text{Ker}(q) = p(V)$; that is

$$0 \to V \to E \to W \to 0$$

is exact. We are always going to assume that this sequence is split as a sequence of R modules. (Notice that this is the case if R is a field F, or if W is a projective R module.)

As an example, we can take $E = V \times W$ with the derivation $D_0((v,w)) = (D_V(v), D_W(w))$ and $p(v) = (v,0)$ and $q((v,w)) = w$. In fact this is almost the general case except we replace D_0 by some other derivation D. All such have the form $D = D_0 + T$ where $T \in \text{Hom}_R(E,E)$ is a linear transformation, which we write as

$$T((v,w)) = (T_{VV}(v) + T_{WV}(w), T_{VW}(v) + T_{WW}(w))$$

where $T_{VV}: V \to V$, $T_{WV}: W \to V$, $T_{VW}: V \to W$, and $T_{WW}: W \to W$ are linear transformations. For p and q to be differential requires that for p, $D(p(v)) = p(D_V(v)$ or that $D((v,0)) =$

$(D_V(v),0)+(T_{VV}(v),T_{VW}(v))=(D_V(v),0)$ which implies that $T_{VV}=0$ and $T_{VW}=0$. For q, we need that $D_W(q(v,w))=q(D(v,w))$ or $D_W(w)=q(D_V(v),D_W(w)+(T_{WV}(w),T_{WW}(w))=D_W(w)+T_{WW}(w)$. Thus $T_{WW}=0$ and we are left with $T(v,w)=(T_{WV}(w),0)$ and $D((v,w))=(D_V(v)+T_{WV}(w),D_W(w))$. Notice that the extensions of W correspond to $\mathrm{Hom}_R(W,V)$.

The extension will be split (differentially isomorphic to $V\times W$) if there is a differential homomorphism $\Phi:W\to E$ such that $q\Phi$ is the identity on W. Such a Φ must have the form $w\mapsto(\phi(w),w)$ for some R linear $\phi:W\to V$. For Φ to be differential we must have $\Phi(D_W(w))=D(\Phi(w))$. or $(\phi(D_W(w)),D_W(w))=(D_V(\phi(w))+T_{WV}(w),D_W(w))$ Thus $\phi D_W=D_V\phi+T_{WV}$, or $T_{WV}=\phi D_W-D_V\phi$, which means in the differential module $\mathrm{Hom}_R(W,V)$, $T_{WV}=D(\phi)$. Conversely, if $\psi:W\to V$ is such that $D(\psi)=T_{WV}$, then $\Psi:W\to E$ by $\Psi(w)=(\psi(w),w)$ is differential and satisfies $q\Psi$ is the identity. We record these calculations is the following:

Proposition 3.7 *Let V and W be differential modules over the differential ring R. Then differential structures D_E on the R module $E=V\times W$ such that the injection p on the first coordinate ($p(v)=(v,0)$) and the projection q on the second coordinate ($q((v,w))=w$) are differential correspond to linear transformations $T\in \mathrm{Hom}_R(W,V)$: $D_E((v,w))=(D_V(v)+T(w),D_W(w))$. There is a differential morphism $\Phi:W\to E$ such that $q\Phi$ is the identity if and only if there is $\phi\in \mathrm{Hom}_R(W,V)$ such that $D(\phi)=T$. If there is such a ϕ then $w\mapsto(\phi(w),w)$ is such a Φ.*

In general, of course, D is not surjective on $\mathrm{Hom}_R(W,V)$. For example, if R is the field $\mathbb{C}(x)$ and $V=W=\mathbb{C}(x)$ then $\mathrm{Hom}_R(V,W)=\mathbb{C}(x)$ and then $T=1/x$ is not a derivative, and hence yields a non-split extension of $\mathbb{C}(x)$ by $\mathbb{C}(x)$. In the previous section we interpreted extensions as elements of Ext^1 and vice versa. Thus we know $\text{Ext}^1_{\mathbb{C}(x)}(\mathbb{C}(x), \mathbb{C}(x))^{\text{diff}} \neq 0$.

On the other hand, for $R = R^{PV}$ (for some differential field F with algebraically closed characteristic zero field of constants C) then D is surjective on R hence on $\text{Hom}_R(R^{(n)}, R^{(m)})$ where the derivations on $R^{(n)}, R^{(m)}$ are coordinate-wise. Thus all extensions of $(R^{(m)}, D_0)$ by $(R^{(n)}, D_0)$ are split. (We are using the fact that $R^{(n)}$ is projective so Proposition 3.7 applies.)

The case $R = R^{PV}$ is special in other ways. Let V be any finitely generated completely D finite-differential R module. Then there is a finite F dimensional F differential submodule V_0 of V which generates V as a differential R module; that is $R \otimes_F V_0 \to V$ by $r \otimes v \mapsto rv$ is surjective. By Theorem 2.6 $R \otimes_F V_0$ has a basis of constants; let $y_1, \ldots, y_n$ be their images under $R \otimes_F V_0 \to V$, which are also constants. We assume that none of the y_i are zero. We consider the submodules $Ry_1 + \cdots + Ry_k$ of V as k varies from 1 to n. After renumbering, we can assume that k is maximal with respect to the sums being direct. If $k = n$ then V is a direct sum of copies of R as a differential R module. If $k < n$, consider Ry_{k+1}. If $(Ry_1 + \ldots Ry_k) \cap Ry_{k+1} = 0$ then the sum $Ry_1 + \cdots + Ry_{k+1}$ is directly contrary to the definition of k. Therefore, the intersection is non-zero. But the intersection is also an R differential submodule of Ry_{k+1}. Since y_{k+1} is a constant, $R \to Ry_{k+1}$ by $r \mapsto ry_{k+1}$ is a differential R module epimorphism. Since R^{PV} is simple as a differential ring, R is a simple differential R module. In particular, the kernel of $R \to Ry_{k+1}$ is zero (it's not R because $1 \mapsto y_{i+1} \neq 0$). Thus Ry_{i+1} is simple, so $(Ry_1 + \ldots Ry_k) \cap Ry_{k+1} = Ry_{k+1}$. Hence $y_{k+1} \in (Ry_1 + \cdots + Ry_k)$. Repeating this argument shows that all y_i belong to the intersection, and thus $V = Ry_1 \oplus \cdots \oplus Ry_k$. Thus:

Proposition 3.8 *Every finitely generated completely D finite R^{PV} module has a finite basis of constants.*

Proposition 3.8 considers a differential ring (R^{PV}) and a differential module (V) both of which are completely D finite F modules. We are going to explore this situation in depth.

3.6 $R \cdot D$ MODULES

Let R be a commutative ring which is a differential F algebra, where F is a differential field of characteristic zero with an algebraically closed field of constants C. Assume that as a differential F module, R is completely D finite. Let V be a differential R module (and hence *a fortiori* a differential F module). If V is completely D finite, then we say V is an $R \cdot D$ module. Generally speaking, there are differential R modules which are not $R \cdot D$ modules; for example, an infinite product of copies of R in most cases.

It is clear that submodules and quotient modules of $R \cdot D$ modules are $R \cdot D$ modules, as are arbitrary direct sums and direct limits. So are tensor products of $R \cdot D$ modules. This is not the case for R homomorphism spaces: consider $\mathrm{Hom}_R(\oplus_i R, R) \cong \Pi_i \mathrm{Hom}_R(R, R)$. However, we do have:

Lemma 3.12 *Let V and W be $R \cdot D$ modules with V finitely generated. Then $Hom_R(V, W)$ is an $R \cdot D$ module.*

Proof. Let V_0 be an F finite-dimensional differential F submodule of V containing a set of generators of V as an $R \cdot D$ module. The surjection $R \otimes_F V_0 \to V$ by $r \otimes v \mapsto rv$ determines a differential R module injection $\mathrm{Hom}_R(V, W) \to \mathrm{Hom}_R(R \otimes V_0, W)$, so it suffices to prove that $\mathrm{Hom}_R(R \otimes V_0, W)$ is an $R \cdot D$ module, which amounts to showing that $\mathrm{Hom}_R(R \otimes V_0, W)$ is completely D finite. This is a statement about differential F modules, and we have the differential F module isomorphism $\mathrm{Hom}_R(R \otimes V_0, W) \to \mathrm{Hom}_F(V_0, W)$, and we have seen above that this latter is completely D finite.

In the proof of Lemma 3.12 we used a special case of the observation that if W is an $F \cdot D$ module then $R \otimes_F W$ is an $R \cdot D$ module via R action on the first factor and differential action $D_R \otimes 1 + 1 \otimes D_W$, the key observation being that since both R and W are completely D finite so is $R \otimes_F W$. This is not usually the case for the adjoint induction $\mathrm{Hom}_F(R, W)$, although again this is a differential R module, differential in the usual way and R action by $rh(s) = h(rs)$ for $h : R \to W$ and $r, s \in R$. The problem is that as an F vector space R may be infinite dimension (in fact usually is) and hence $\mathrm{Hom}_F(R, W) \cong \Pi_1^\infty \mathrm{Hom}_F(F, W)$ is not completely D finite. We can possibly remedy this situation by considering the maximal D finite submodule $\mathrm{Hom}_F(R, W)^{\mathrm{fin}}$. Note that if V is an $R \cdot D$ module then $\mathrm{Hom}_R^{\mathrm{diff}}(V, \mathrm{Hom}_F(R, W)) = \mathrm{Hom}_R^{\mathrm{diff}}(V, \mathrm{Hom}_F(R, W)^{\mathrm{diff}})$. Then using adjoint associativity, we have for an $R \cdot D$ module V and an $F \cdot D$ module V

$$\mathrm{Hom}_R^{\mathrm{diff}}(V, \mathrm{Hom}_F(R, W)^{\mathrm{fin}}) \cong \mathrm{Hom}_R^{\mathrm{diff}}(V, \mathrm{Hom}_F(R, W))$$

and

$$\mathrm{Hom}_R^{\mathrm{diff}}(V, \mathrm{Hom}_F(R, W)) \cong \mathrm{Hom}_F^{\mathrm{diff}}(V \otimes_R R, W)$$

then using the identification $V \otimes_R R \cong V$, the final "Hom" is $\mathrm{Hom}_F^{\mathrm{diff}}(V, W)$ where V is regarded as an $F \cdot D$ module. To be explicit, if $h : V \to W$ is an F differential homomorphism where V is an $R \cdot D$ module and W is an $F \cdot D$ module then the corresponding $R \cdot D$ morphism $H : V \to \mathrm{Hom}_F(R, W)$ is given by $H(v)(r) = h(rv)$. So, for example, v is in the kernel of H means $h(Rv) = 0$. In particular, if h is monic so is H. This leads us to the following construction of injective $R \cdot D$ modules:

Theorem 3.5 *Let I be an injective $F \cdot D$ module. Then $Hom_F(R, I)^{fin}$ is an injective $R \cdot D$ module.*

Proof. As a functor on $R \cdot D$ modules $\mathrm{Hom}_R(\circ, \mathrm{Hom}_F(R, I)^{\mathrm{fin}}$ is isomorphic to $\mathrm{Hom}_F(\circ, I)$; since this latter is exact, so is the former.

If we apply Theorem 3.5 to the case $R = R^{PV}$ and $I = R^{PV}$, we see that $\mathrm{End}_F(R^{PV})^{\mathrm{fin}}$ is an injective $R \cdot D$ module. Let $\iota \in \mathrm{End}_F(R^{PV})^{\mathrm{fin}}$ be the identity ($D(\iota) = 0$ so ι is D finite), and then $R^{PV} \to \mathrm{End}_F(R^{PV})^{\mathrm{fin}}$ by $r \mapsto r\iota$ is an $R \cdot D$ embedding of R^{PV} into an injective module. Since by Proposition 3.8 any finitely generated R^{PV} module V is a finite dierect sum (say n) of copies of R^{PV} $V \to (\mathrm{End}_F(R^{PV})^{\mathrm{fin}})^{(n)}$ is an embedding of V in an injective module.

As a further consequence of Theorem 3.5 and the remarks preceding it, we have the following regarding embedding $R \cdot D$ modules in injective modules:

Corollary 3.13 *Let V be an $R \cdot D$ module and let $h : V \to I$ be an $F \cdot D$ embedding of V in an injective $F \cdot D$ module. Then $H : V \to Hom_F(R, I)^{fin}$ by $H(v)(r) = h(rv)$ is an embedding of V on an injective $R \cdot D$ module.*

We know that, for any $F \cdot D$ module V, $V \otimes_F R^{PV}$ is an injective $F \cdot D$ module and $h : V \to V \otimes_F R^{PV}$ by $h(v) = v \otimes 1$ is an $F \cdot D$ embedding. In the context of Corollary 3.13 this means that $H : V \to \operatorname{Hom}_F(R, V \otimes_F R^{PV})$ by $H(v)(r) = rv \otimes 1$ is an $R \cdot D$ embedding of V into an injective $R \cdot D$ module. Note that this construction is natural in V.

It is standard category theory that the category of $R \cdot D$ modules has enough injectives. Corollary 3.13 shows that it has enough injectives of the form $\operatorname{Hom}_F(R, I)^{\text{fin}}$. In either case we can form injective resolutions and hence do homological algebra. In particular, we can form the right derived functors $R^i\operatorname{Hom}_R^{\text{diff}}(V, \circ)$ where V is an $R \cdot D$ module and the domain of the functor is the category of $R \cdot D$ modules.

We adopt the following notation:

$$R^i\operatorname{Hom}_R^{\text{diff}}(V, \circ) = \operatorname{Ext}^i_{R \cdot D}(V, \circ).$$

Thus for $R \cdot D$ modules V and W $\operatorname{Ext}^0_{R \cdot D}(V, W) = \operatorname{Hom}_R^{\text{diff}}(V, W)$. If $W \to E$ is an embedding of the $R \cdot D$ module W into an injective $R \cdot D$ module E then the long exact sequence of $\operatorname{Ext}_{R \cdot D}(V, \circ)$ associated to the short exact sequence $0 \to W \to E \to E/V \to 0$ gives an isomorphism

$$\operatorname{Hom}_R^{\text{diff}}(V, E/W)/\operatorname{Hom}_R^{\text{diff}}(V, E) \cong \operatorname{Ext}^1_{R \cdot D}(V, W).$$

Exactly as we saw in the case of $F \cdot D$ modules, if we start with an $R \cdot D$ homomorphism $f : V \to E/W$ and let $X \subset E$ be the inverse image of $f(V)$ under the projectione $E \to E/W$ then we have a short exact sequence $0 \to V \to X \to W \to 0$ and thus we can interpret elements of $\operatorname{Ext}^1_{R \cdot D}(V, W)$ as short exact sequences. If W happens to be projective as an R module so that X as an R module is $V \times W$ then we can further analyze the short exact sequence in terms of elements of $\operatorname{Hom}_R(W, V)$. As we now show, this happens in a broad class of differential rings.

3.7 SIMPLE COMPLETELY D FINITE RINGS AND MODULES

By construction, Picard–Vessiot rings are simple: they have no proper differential ideals. They are also completely D finite. In this section, we consider rings. To begin with, we suppose R is a simple differential ring. Since 0 is a maximal differential ideal of R, and maximal differential ideals are prime ideals, 0 is a prime ideal of R so R is an integral domain. We let K denote the quotient field of R and S the set of non-zero elements of R so $K = S^{-1}R$. If M is any R module we have an R homomorphism $M \to S^{-1}M = K \otimes_R M$ by $m \mapsto m/1 = 1 \otimes m$; if M is a differential module this is a differential homomorphism. If $m/1 = 0$, this means there is an $s \in S$ such that $sm = 0$. If M is finitely generated by $m_1, \ldots, m_k$ and $m_1/1 = \cdots = m_k/1 = 0$ then there are $s_1, \ldots, s_k$ in S such that $s_i m_i = 0$ all i. Let $s_0 = s_1 s_2 \ldots s_k$; then $s_0 m_i = 0$ all i so $sm = 0$ for all $m \in M$. For any module M we let $\text{Ann}(M) = \{r \in R | rm = 0 \text{ for all } m \in M\}$ (the *annihilator* of M). The annihilator of M is an ideal of R. If M is a differential module, then the annihilator is a differential ideal: if $r \in \text{Ann}(M)$ then for $m \in M$ $D(r)m = D(rm) - rD(m)$ and both terms on the right of this equation are zero, $D(rm) = D(0) = 0$ since $rm = 0$, and $rD(m) = 0$ because $r \in \text{Ann}(M)$ and $D(m) \in M$. If R is differentially simple then the differential ideal $\text{Ann}(M)$ is zero (except in the trivial case $M = 0$).

We saw above that if M is finitely generated and $S^{-1}M = 0$ then $\text{Ann}(M) \neq 0$ so $M = 0$. Since the functor $S^{-1}(\circ)$ is exact and in particular preserves inclusions, and since every differential module is a directed union of its finitely generated differential submodules, we have this conclusion ($S^{-1}M = 0$ implies $M = 0$) for all differential R modules

M. Again since $S^{-1}(\circ)$ is exact, (so preserves kernels and cokernels) if $f : M \to N$ is a morphism of differential modules then f has zero kernel (respectively zero cokernel) if and only if $S^{-1}f$ does. Here is an important application of these observations:

Theorem 3.6 *Let R be a simple differential ring which is an F algebra, and let M be a differential R module generated by a finite F dimensional differential submodule module. Then M is projective as an R module.*

Proof. Let V be an F finite-dimensional differential F submodule of M containing a set of generators, so that $f : R \otimes_F V \to M$ by $r \otimes v \mapsto rv$ is a surjective differential morphism. Consider the induced morphism $\phi : \mathrm{Hom}_R(M, R \otimes V) \to \mathrm{Hom}_R(M, M)$. When $S^{-1}(\circ)$ is applied, $S^{-1}\phi$ becomes surjective because the K vector space surjection $K \otimes V \to S^{-1}M$ has a K section. But then ϕ is surjective, which means the R module homomorphism $f : R \otimes_F V \to M$ has an R module section, so that M is isomorphic to an R module direct summand of the free R module $R \otimes_F V$. Thus M is R projective.

The hypotheses of Theorem 3.6 are satisfied when R is a simple completely D finite-differential F algebra and M is a finitely generated $R \cdot D$ module. Here are some consequences in that case:

Corollary 3.14 *Let R be a simple completely D finite-differential F algebra. Let M be a finitely generated $R \cdot D$ module and let $N \subseteq M$ be a differential R submodule. Then N is an R module direct summand of M. If follows that finitely generated $R \cdot D$ modules have the* d.c.c *and in particular non-zero socles over which they are essential.*

Proof. Since M/N is a finitely generated $R \cdot D$ module it is R projective, so the R surjection $M \to M/N$ splits as R modules displaying both M/N and N as R direct summand of M. The differential submodules of M are thus all R direct summands; they are all then projective of rank bounded by that of M, which shows that M has the descending chain condition. The assertions about socles then follows.

A further consequence of Theorem 3.6 is that a short exact sequence $0 \to V \to X \to W \to 0$ of finitely generated $R \cdot D$ modules splits as R modules. Thus we can apply the calculations with such modules done above. In particular, we know that the sequence can be described by an R linear map ϕ in $\mathrm{Hom}_R(W, V)$, and the sequence is split if and only if ϕ is a derivative. Combined with the interpretation of Ext as short exact sequences we get the description

$$\mathrm{Hom}_R(W, V)/D(\mathrm{Hom}_R(W, V)) \cong \mathrm{Ext}^1_{R \cdot D}(V, W).$$

The completely D finite-differential ring R^{PV} is differentially simple, so all its finitely generated completely D finite modules are R^{PV} projective. Of course, we know a lot more about R^{PV}. For example, by Proposition 3.8 we know that finitely generated completely D finite R^{PV} are actually free as R^{PV} modules, indeed are differentially isomorphic to a finite sum of copies of R^{PV}, We can express this by saying: Let V be a finitely generated completely D finite R^{PV} module. Then V^D is C finite dimensional and $R^{PV} \otimes_C V^D \to V$ is an R^{PV} differential isomorphism. In this form, it applies to all completely D finite differential R^{PV} modules:

Corollary 3.15 *(of Proposition 3.8) Let V be an $R^{PV} \cdot D$ module. Then $R^{PV} \otimes_C V^D \to V$ by $r \otimes v \mapsto rv$ is an isomorphism of $R^{PV} \cdot D$ modules.*

Proof. Suppose $x = \sum_i r_i \otimes v_i$ lies in the kernel. There is a finitely generated $R^{PV} \cdot D$ submodule W of V which contains all the v_i's, so all the v_i's are in W^D. Since $R^{PV} \otimes_C W^D \to W$ is injective, $x = 0$. Thus the map is injective. If $v \in V$ then there is a finitely generated $R^{PV} \cdot D$ submodule U with $v \in U$ and then v lies in the image of the isomorphism $R^{PV} \otimes U^D \to U$, say $y = \sum_j r_j \otimes u_j \mapsto v$. Since each $u_j \in V^D$ also, we have that $y \in R^{PV} \otimes_C V^D$ and $y \mapsto v$. Thus the map is also onto.

An R^{PV} differential homomorphism $f : V \to W$ of $R^{PV} \cdot D$ modules carries V^D to W^D so that the diagram

$$\begin{array}{ccc} R^{PV} \otimes_C V^D & \xrightarrow{1 \otimes f|V^D} & R^{PV} \otimes_C W^D \\ \downarrow & & \downarrow \\ V & \xrightarrow{f} & W, \end{array}$$

where the vertical maps are those of Corollary 3.15, commutes.

It follows that $R^{PV} \otimes_C (\circ)$ from C vector spaces to $R^{PV} \cdot D$ modules is an equivalence, indeed an isomorphism, of categories. From which we conclude:

Every $R^{PV} \cdot D$ module, finitely generated or not, is a differential direct sum of copies of R^{PV}. We also know that, since all C vector spaces are injective, all $R^{PV} \cdot D$ modules are injective; by the previous observation, this is equivalent to noting that since C is an injective vector space $R^{PV} = R^{PV} \otimes_C C$ is an injective $R^{PV} \cdot D$ module by the category isomorphism.

For any simple completely D finite differential ring R, Theorem 3.6 says that finitely generated $R \cdot D$ modules are R projective. Since all $R \cdot D$ modules are direct limits of finitely generated ones, this implies that every $R \cdot D$ module is a direct limit of R projectives, and hence R flat.

The Picard–Vessiot ring R_L associated to a monic linear homogeneous differential operator L over F is completely D finite and differentially simple. It is also a subring of R^{PV}. Let V be an $R_L \cdot D$ module, and consider $\mathrm{Hom}_{R_L}^{\mathrm{diff}}(V, R^{PV})$. Since $\mathrm{Hom}_{R_L}^{\mathrm{diff}}(V, \circ) \cong \mathrm{Hom}_{R^{PV}}^{\mathrm{diff}}(R^{PV} \otimes_R V, \circ)$. as a functor of $R_L \cdot D$ modules $\mathrm{Hom}_{R_L}^{\mathrm{diff}}(\circ, R^{PV})$ can be written as a composition of the functors $(\circ) \mapsto R^{PV} \otimes_{R_L} (\circ)$ and $(\circ) \mapsto \mathrm{Hom}_{R^{PV}}^{\mathrm{diff}}(\circ, R^{PV})$ both of which are exact (the former since R^{PV} is R_L flat, as are all $R_L \cdot D$ modules). It follows that R^{PV} is an injective R_L module. And if V is any $R_L \cdot D$ module then $R^{PV} \otimes_{R_L} V$, which, as an R^{PV} module (hence as an R_L module), is a direct sum of copies of R^{PV}, is also an injective R_L module. Thus the $R_L \cdot D$ module map $V \to R^{PV} \otimes_{R_L} V$ by $v \mapsto 1 \otimes v$ is an embedding of V into an $R_L \cdot D$ injective. We prove this last assertion, which actually is part of a more general observation:

Lemma 3.13 *Let $R_1 \subset R_2$ be simple completely D finite differential commutative differential F algebras. Then the functor $R_2 \otimes_{R_1} (\circ)$ from $R_1 \cdot D$ modules to $R_2 \cdot D$ modules is exact and faithful, and $(\circ) \to R_2 \otimes_{R_1} (\circ)$ by $x \mapsto x \otimes 1$ is injective as a map of $R_1 \cdot D$ modules.*

Proof. Because R_2 is an $R_1 \cdot D$ module, it is flat, so $R_2 \otimes_{R_1} (\circ)$ from $R_1 \cdot D$ is exact. Let V be an $R_1 \cdot D$ module. Since V is flat, tensoring with it carries monomorphisms to monomorphisms. In particular, $R_1 \otimes_{R_1} V \to R_2 \otimes_{R_1} V$ is a monomorphism, which, since the domain is isomorphic to

V, is the final assertion. Thus $R_2 \otimes_{R_1} V = 0$ implies $V = 1 \otimes V = 0$, so the functor is faithful.

Note that Lemma 3.13 does not say that R_2 is a faithfully flat R_1 module. We are going to explore this idea. We begin with a discussion of $R \cdot D$ modules for any simple completely D finite-differential commutative differential F algebra R. Suppose V is a finitely generated $R \cdot D$ module. We can find a finite-dimensional differential F module $V_1 \subset V$ such that $p : R \otimes_F V_1 \to V$ by $p(r \otimes v) = rv$ is an $R \cdot D$ surjection. Let K be the kernel of p. Since V is finitely generated projective as an R module, K is an R module direct summand of $R \otimes_F V_1$, and therefore applying R duality to the exact sequence

$$0 \to K \to R \otimes_F V_1 \to V \to 0$$

yields an exact sequence of $R \cdot D$ modules

$$0 \to \mathrm{Hom}_R(V, R) \to \mathrm{Hom}_R(R \otimes_F V_1, R) \to \mathrm{Hom}_R(K, R) \to 0.$$

The $R \cdot D$ module K is an R module direct summand of the finitely generated R module $R \otimes_F V_1$, hence K is finitely generated as an R module hence also as $R \cdot D$ module, so there is a finite dimensional differential F submodule $V_2 \subset K$ such that $q : R \otimes_F V_2 \to K$ by $q(r \otimes y) = ry$ is an $R \cdot D$ surjection. This implies that $\mathrm{Hom}_R(K, R) \to \mathrm{Hom}_R(R \otimes_F V_2, R)$ is an injection. Splicing this onto the end of our previous sequence then gives the exact sequence of $R \cdot D$ modules

$$0 \to \mathrm{Hom}_R(V, R) \to \mathrm{Hom}_R(R \otimes_F V_1, R) \to \mathrm{Hom}_R(R \otimes_F V_2, R).$$

Now $\mathrm{Hom}_R(R \otimes_F V_i, R) = \mathrm{Hom}_F(V_i, R)$ as differential modules and as R modules (where R acts on homomorphism on the right by $r \cdot f(x) =$

$rf(x)$) compatibly so as $R \cdot D$ modules. Thus finally our exact sequence is

$$0 \to \mathrm{Hom}_R(V, R) \to \mathrm{Hom}_F(V_1, R) \to \mathrm{Hom}_F(V_2, R).$$

Since V is a finitely generated projective R module, $V \cong \mathrm{Hom}_R(V, R)$. Also since V_i is a finite-dimensional F vector space, $\mathrm{Hom}_F(V_i, R) \cong \mathrm{Hom}_F(V_i, F) \otimes_F R$. Let $W = \mathrm{Hom}_F(V, F)$ and let $W_i = \mathrm{Hom}_F(V_i, F)$. Then W can be any finitely generated $R \cdot D$ module, and the above exact sequence becomes

$$0 \to W \to R \otimes_F W_1 \to R \otimes_F W_2.$$

As an R module, $R \otimes_F W_1$ (respectively $R \otimes_F W_2$) is free of rank $n = \dim_F(W_1)$ (respectively $m = \dim_F(W_2)$) and hence of the form (R^n, D_A) (respectively (R^m, D_B)) where A (respectively B) is an $n \times n$ (respectively $m \times m$) matrix. Because of the tensor product structure, A and B can be taken to have entries in F. The differential homomorphism $(R^n, D_A) \to (R^M, D_B)$ is then of course given by an $m \times n$ matrix T with entries in R satisfying $T' = TA - BT$; of course, T need not have its entries in F.

Given a W, or its dual V, there are always W_1, W_2, V_1, V_2, to express W as a kernel, or a cokernel. Now suppose R is a Picard–Vessiot ring and V is a finitely generated $R \cdot D$ module with associated V_1 and V_2. Let $S \supseteq R$ be a PIcard–Vessiot ring such that $\mathrm{Hom}_F^{\mathrm{diff}}(V_1, S) = \mathrm{Hom}_F^{\mathrm{diff}}(V_1, R^{PV})$ so that $S \otimes_R V_1$ has a (finite) basis of constants as a differential S module. The image of this basis in $S \otimes_R V$ is then a finite generating set for $S \otimes_R V$ consisting of constants. Using the fact that S is simple, and repeating the argument that led to Proposition 3.8, then show that $S \otimes_F V$ has a basis of constants. We summarize this discussion in the following result:

Proposition 3.9 *Let R be a Picard–Vessiot ring over F. Let V be a finitely generated $R \cdot D$ module. Then there are finite-dimensional differential F modules V_0 and W_0 and $F \cdot D$ morphisms $R \otimes_F V_0 \to V$ (surjective) and $V \to F \otimes_F W_0$ (injective). Let $S \supseteq R$ be the Picard–Vessiot ring for a Picard–Vessiot extension of F splitting V_0. Then $S \otimes_F V$ has a basis of constants as a differential S module. The same is true if S is a Picard–Vessiot ring splitting W_0.*

CHAPTER 4

Differential Projective Modules

4.1 DERIVATIONS OF PROJECTIVE MODULES

When R is a completely D finite-differential commutative algebra over a differential field, where the latter is assumed to be of characteristic zero and has an algebraically closed field of constants, and furthermore R has no proper differential ideals, we have seen that every finitely generated differential R module is projective as an R module. The lengthy string of conditions assumed of R clearly only apply in quite special cases, but those cases include Picard–Vessiot rings associated to linear differential equations over the given field, which arguably is the base case of differential rings and differential modules. In this chapter we are going to look beyond both the D finite case (we will consider differential rings and modules where not every element is the solution of a linear differential equation) and the simple case (we will permit R to have proper ideals).

DOI: 10.1201/9781003452164-4

However, we are going to conserve the property, automatically in the D finite simple case, that the differential R modules, at least the finitely generated ones, are R projective.

There are a variety of reasons for this restriction. Perhaps the most important is that projective modules are a natural extension of the case of free modules. A recurring theme has been the differential modules of the form $M = (R^n, D_A)$ where A is an $n \times n$ matrix over R. As we shall see, any R finitely generated differential module is a homomorphic image of such a module, which is hence the quotient of such a differential module by a differential submodule. Any differential submodule may occur, of course, but as a first step we can look at those which are R module direct summands. These are, of course, projective.

In this chapter, R always denotes a commutative differential ring. Generally speaking, we intend to look at differential modules which are finitely generated as R modules. To order the adjectives properly, we call such a differential module a *differential finitely generated* R module. This would be in distinction from a finitely generated differential module, which is a differential module (like $R[X; D]$) which is finitely generated (by $X^0 = 1$) as a differential module, but as an R module is not necessarily finitely generated. We have previously followed this convention when we studied differential free modules (differential modules which are free as R modules; the ones which are finitely generated as R modules are the (R^n, D_A)'s) as opposed to free differential modules, again like $R[X : D]$ which is free on $X^0 = 1$.

As an example of the properties of differential finitely generated modules, we have the following.

Proposition 4.1 *Let M be a differential finitely generated R module. Then there is an n, an $n \times n$ matrix A with entries in R and a surjective differential R linear homomorphism $f : (R^n, D_A) \to M$.*

Proof. Let $x_1, \ldots, x_n$ be a set of generators of M as an R module. Suppose for each i that $D(x_i) = \sum_j a_{ij} x_j$. Let $A = [a_{ji}]$. Let $f : R^n \to M$ be the unique R module homomorphism such that $f(e_i) = x_i$. Then f is a surjective differential R module homomorphism $(R^n, D_A) \to M$.

We can also prove Proposition 4.1 using Proposition 1.8.

Recall that $R[\epsilon]$ with $\epsilon^2 = 0$ is an R algebra via $r \mapsto r + D_R(r)\epsilon$ so that for any R module M, the $R[\epsilon]$ module $M[\epsilon] = R[\epsilon] \otimes_R M$ is an R module via $r \cdot)(s + t\epsilon) = (r + D_R(r)\epsilon)(s + t\epsilon) = rs + (rt + D_R(r)s)\epsilon$.

For the differential module M, $M \to M[\epsilon]$ by $m \mapsto m + D_M(m)\epsilon$ is an R module homomorphism. Let F be a free R module and $f : F \to M$ be an R module surjection. Then we have an $R[\epsilon]$ module surjection $F[\epsilon] \to M[\epsilon]$ which is also an R module surjection when we use $R \to R[\epsilon]$.

Consider the diagram

$$\begin{array}{ccc} F & & F[\epsilon] \\ {\scriptstyle f}\downarrow & & \downarrow \\ M & \longrightarrow & M[\epsilon] \end{array}$$

Because the right vertical arrow is surjective and F is free, there is a map $F \to F[\epsilon]$ making the diagram commutative:

$$\begin{array}{ccc} F & \xrightarrow{h} & F[\epsilon] \\ {\scriptstyle f}\downarrow & & \downarrow \\ M & \longrightarrow & M[\epsilon]. \end{array}$$

We would like h to give a derivation on F so that f is differential.

That is, we want $h(x) = x + D_F(x)\epsilon$ for $x \in F$. Unfortunately, all the commutativity of the diagram guarantees is that $h(x) = p(x) + q(x)\epsilon$ where $f(p(x)) = f(x)$. To remedy this, we consider instead the module $F + M\epsilon \cong F[\epsilon]/\text{Ker}(f)\epsilon$. As an $R[\epsilon]$ module R acts on $F + M\epsilon$ on each summand, but $\epsilon \cdot (x + m\epsilon) = f(x)\epsilon$.

Since $\text{Ker}(f)\epsilon$ goes to 0 under $F[\epsilon] \to M[\epsilon]$ the surjection of differential R modules $F[\epsilon] \to M[\epsilon]$ factors as a differential R module surjection $F[\epsilon] \to F + M\epsilon$ followed by $F + M\epsilon \to M[\epsilon]$ which we can use in the partial diagram

$$\begin{array}{ccc} F & & F[\epsilon] \\ \downarrow & & \downarrow \\ F & \xrightarrow[\phi]{} & F + M\epsilon \end{array}$$

where $\phi(x) = x + D_M(f(x))\epsilon$. We note that $\phi(rx) = rx + (D_R(r)f(x) + rD_M(f(x)))\epsilon = (r + D_R(r)\epsilon)(x + D_M(f(x))\epsilon$ which shows that ϕ is R linear. Note also that

$$\begin{array}{ccc} F & \xrightarrow{\phi} & F + M[\epsilon] \\ f\downarrow & & \downarrow \\ M & \longrightarrow & M[\epsilon] \end{array}$$

commutes.

Now, because $F[\epsilon] \to F + M\epsilon$ is surjective and F is free, we have a map $\Phi : F \to F[\epsilon]$ such that

$$\begin{array}{ccc} F & \xrightarrow{\Phi} & F[\epsilon] \\ \downarrow & & \downarrow \\ F & \xrightarrow[\phi]{} & F + M\epsilon \end{array}$$

commutes. Since $\phi(x) = x + D_M(f(x))$ and is R linear, $\Phi(x) = x + D_F(x)\epsilon$ where $D_F : F \to F$ is a derivation. And putting both diagrams together we have a commutative diagram

$$\begin{array}{ccc} F & \xrightarrow{\Phi} & F[\epsilon] \\ \downarrow & & \downarrow \\ M & \longrightarrow & M[\epsilon] \end{array}$$

which shows that $f : (F, D_F) \to (M, D_M)$ is now a differential R module homomorphism. This is actually an improvement on Proposition 4.1 because we no longer have to assume M is finitely generated as an R module.

We note what was actually used in this argument: we postulated an R module surjection $F \to M$, and we used the fact that F was free to lift the morphism $F \to F + M\epsilon$ to the source of the surjection $F[\epsilon] \to F + M\epsilon$. For lifts through surjectives like this, we only need that F is projective. Thus our argument proves the following:

Proposition 4.2 *Let M be a differential R module and P a projective R module. Suppose $f : P \to M$ is an R module surjection. Then there is a derivation of P such that $f : (P, D_P) \to (M, D_M)$.*

Since 0 is a differential R module, for any projective R module P, Proposition 4.2 applied to $P \to 0$ shows that there is a differential structure on P. We observe this as a corollary:

Corollary 4.1 *Every projective R module has a differential structure.*

It is worthwhile noticing how proof of the Corollary, which is the proof of Proposition 4.2, works in the case $M = 0$: the map $\phi : P \to$

$P + M\epsilon]$ is just the identity, and the surjection $P[\epsilon] \to P + M\epsilon$ is just the surjection $P[\epsilon] \to P$. Hence $\Phi : P \to P[\epsilon]$ is just the completion of the diagram

$$\begin{array}{ccc} P & & P[\epsilon] \\ \downarrow & & \downarrow \\ P & \xrightarrow[\phi]{} & P \end{array}$$

which amounts to lifting the identity from P to $P[\epsilon]$; in other words, splitting the surjection $P[\epsilon] \to P$, which happens because P is projective.

One can ask whether there is an analog of Corollary 4.1 for other classes of modules, for instance, does every R module have a differential structure? That depends on the ring: we know that for a Picard–Vessiot ring R differential finitely generated modules are projective, and then with Corollary 4.1 that projective modules are differential. Unless the ring is a field, however, there are always proper ideals I, and then R/I is a finitely generated R module which is not projective, and hence can't carry a differential structure.

4.2 IDEMPOTENT MATRICES AND PROJECTIVE COORDINATE SYSTEMS

Returning to the general R, there is an additional "constructive" method to produce differential structures on projective modules, at least if the module is explicitly given. By that we mean the following: a finitely generated projective module P, say with n generators, over a commutative ring R , is a direct summand of a free R module of rank n. That means we can regard $P \subseteq R^n$ and there is a submodule $Q \subseteq R^n$ such that

$P \cap Q = 0$ and $P + Q = R^n$. Any element $x \in R^n$ can be written uniquely as a sum $x_P + x_Q$ where $x_P \in P$ and $x_Q \in Q$. Then $E : R^n \to R^n$ by $x \mapsto x_P + 0$ is an R linear endomorphism of R^n whose image is P. Also $E^2 = E$. Conversely, given an idempotent R linear endomorphism T of R^n we have that $T(R^n)$ is a direct summand of R^n (the complementary summand is $(I - T)(R^n)$ and hence projective. Using the standard basis of R^n identifies $\text{End}_R(R^n)$ with $M_n(R)$. We call a finitely generated projective R module *explicitly given* if for some an idempotent matrix $e \in M_n(R)$ is given with $eR^n \cong P$.

If R is a differential ring, so is $M_n(R)$, using coordinate-wise differentiation of the entries of a matrix: $[a_{ij}]' = [a'_{ij}]$. We have the following formula regarding the derivative of an idempotent element of a (not necessarily commutative) ring (we use the notation that $[a, b] = ab - ba$ for elements a and b of a ring):

Lemma 4.1 *Let A be a not necessarily commutative differential ring. Let $e \in A$ be an idempotent. Then $e' + [[e, e'], e] = 0$.*

Proof. Since $e = e^2$, $e' = ee' + e'e$. Now multiply this equation on the left and right by e to get $ee'e = ee'e + ee'e$ which implies that $ee'e = 0$. Then we expand $[[e, e'], e]$ to get $[ee' - e'e, e] = ee'e - e'e - ee'ee'e$. Since $ee'e = 0$, this reduces to $-(ee' + e'e)$ which we recognize as $-e'$. Therefore, $e' + [[e, e'], e] = 0$.

We apply Lemma 4.1 to the case of the idempotent matrix $e \in M_n(R)$ whose image is the projective module P. Consider the differential R module $M = (R^n, D_{[e,e']})$. For $x \in M$, $D_M(ex) = (ex)' + [e, e']ex = e'x + ex' + [e, e']ex$ while $eD_M(x) = e(x' + [e, e']x) = ex' + e[e, e']x$. Thus $D_M(ex) - eD_M(x) = e'x + [e, e']ex - e[e, e']x$. This right-hand side of this last equality simplifies to $e'x + [[e,e'],e]x$ which is zero by Lemma 4.1. Thus $D_M(ex) = eD_M(x)$, proving that e is a differential endomorphism of M. That makes its image, P, a differential submodule. Since e is differential, so is $I_n - e$, which makes its image (call it Q) a differential submodule as well. Moreover, as differential modules, we have $P \oplus Q = M$.

In the above, e is external to P, in that it comes from a choice of R module surjection $R^n \to P$. That surjection, and its splitting which shows that P is projective, can be encoded in data that is internal to P, that is using the associated projective coordinate system. Thus we could equivalently start with the latter. Thus let P be an R module possessing a projective coordinate system x_i, f_i, $1 \leq i \leq n$ where the x_i are in P and the f_i are in $\mathrm{Hom}_R(P,R)$ such that for all $x \in P$, $x = \sum_i f_i(x)x_i$. Note that this implies that $p : R^n \to P$ by $p([a_i,\ldots,a_n]^T) = \sum_i a_i x_i$ is surjective and that $q : P \to R^n$ by $q(x) = [f_1(x),\ldots,f_n(x)]^T$ satisfies $p(q(x)) = x$: that is q splits p implying that P is projective. Consider $e = q \circ p : R^n \to R^n$, explicitly $e([a_n,\ldots,a_n]^T) = [f_1(\sum_i a_i x_i),\ldots,f_n(\sum_i a_a x_i)]^T$. Then e applied to the jth standard basis vector $[0,\ldots,1,\ldots 0]^T$ is $[f_1(x_j),\ldots,f_n(x_j)]^T$ which means that the jth column of the matrix of e in the standard basis (which we also call e) is $[f_1(x_j),\ldots,f_n(x_j)]^T$. Thus $e = [f_i(x_j)]$. Since x_i, f_i is a projective coordinate system, $x_j = \sum_k f_k(x_j)$ so $f_i(x_j) = f_i(\sum_k f_k(x_j)x_k) = \sum_k f_k(x_j)f_i(x_k) = \sum_k f_i(x_k)f_k(x_j)$ showing that $e^2 = e$.

Let $\alpha = [a_1,\ldots,a_n]^T \in R^n$. Then $e\alpha = [\sum_j a_j f_1(x_j),\ldots,\sum_j a_j f_n(x_j)]^T$ and $\sum_j a_j f_i(x_j) = f_i(\sum_j a_j x_j)$.

Hence $p(e\alpha) = \sum_i \sum_j f_i(\sum_j (a_j x_j))x_i = \sum_i f_i(\sum_j a_j x_j)x_i = \sum_j a_j x_j = p(\alpha)$.

The same formula shows that $q(\sum_j a_j x_j) = [f_1(\sum_j a_j x_j), \ldots, f_n(\sum_j a_j x_j)]^T = e\alpha$ (this is just the explicit calculation that $q \circ p$ is the identity). Now consider p restricted to eR^n. Let $x \in P$. Since p is surjective, we can find $\alpha \in R^n$ such that $x = p(\alpha)$. Since $p(\alpha) = p(e\alpha)$, this means that the restriction of p to eR^n is also surjective. Now suppose $p(e\alpha) = 0$. Then $q(p(e\alpha)) = 0$, so $e\alpha = 0$. Thus p restricted to eR^n is injective and hence gives an isomorphism $eR^n \cong P$.

All of the remarks in the preceding paragraph are comments amount projective coordinate systems and apply to projective modules regardless of whether the ambient ring R is differential. However, as the discussion in the penultimate paragraph explains, when R is a differential ring then the matrix e is a differential endomorphism of $(R^n, D_{[e,e']})$ and hence eR^n is a differential submodule (indeed, direct summand), hence so is $P \cong eR^n$. The derivation on P is given by $D_P(x) = p(D_{[e,e']}(q(x)))$. To write this explicitly, we use that $e' = [f_i(x_j)']$ so that $[e,e'] = ee' - e'e = [\sum_k (f_i(x_k) f_k(x_j)' - f_i(x_k)' f_k(x_j))]$.Thus

$$D_P(x) = y_1 x_1 + \cdots + y_n x_n \text{ where}$$

$$y_i = f_i(x)' + \sum_j f_j(x) (\sum_k (f_i(x_k) f_k(x_j)' - f_i(x_k)' f_k(x_j))$$

Note that this gives an explicit formula for a derivation of P in terms of the given projective coordinate system.

We record these results as a theorem:

Theorem 4.1 *Let P be an R module and suppose that x_i, f_i, $1 \le i \le n$ is a projective coordinate system for P. Let e be the $n \times n$ matrix $[f_i(x_j)]$. Then e is a differential endomorphism of $(R^n, D_{[e,e']})$. In particular, eR^n*

carries the structure of a differential module hence so does P where

$$D_P(x) = y_1x_1 + \cdots + y_nx_n \text{ and}$$
$$y_i = f_i(x)' + \sum_j f_j(x)(\sum_k (f_i(x_k)f_k(x_j)' - f_i(x_k)'f_k(x_j))$$

Theorem 4.1 applies to any finitely generated projective R module $\mathcal{P}$, including one that already has a differential structure $D_{\mathcal{P}}$. To create a projective coordinate system for $\mathcal{P}$ that respects this structure, we start with a projective R module $\mathcal{Q}$ chosen such that $\mathcal{P} \oplus \mathcal{Q}$ has an R basis, say $w_1, \ldots, w_m$. Since $\mathcal{Q}$ is a finitely generated projective R module, Theorem 4.1 guarantees that it carries a differential structure, say $D_{\mathcal{Q}}$ and hence $V = \mathcal{P} \oplus \mathcal{Q}$ has the differential structure $D_V = D_{\mathcal{P}} + D_{\mathcal{Q}}$. Any $w \in V$ can be uniquely written as $w = y + z$ where $y \in \mathcal{P}$ and $z \in \mathcal{Q}$. Then the projection $V \to \mathcal{P}$ by $w = y + z \mapsto y$ and the injection $\mathcal{P} \to V$ by $y \mapsto y + 0$ are both differential homomorphisms.

We write $w_i = y_i + z_i$, $y_i \in \mathcal{P}$ and $z_i \in \mathcal{Q}$. Then if $w \in V$ we have $w = \sum_i r_i z_i = (\sum_i r_i y_i) + (\sum_i r_i z_i)$ with $y = \sum_i r_i y_i \in \mathcal{P}$ and $w = \sum_i r_i z_i \in \mathcal{Q}$. If $w \in \mathcal{P}$ then $w = y = \sum_i r_i y_i$ which shows that the y_i span $\mathcal{P}$. Let $g_i : \mathcal{P} \to R$ be such that $y = \sum_i g_i(y) w_i$, which as we just saw implies that $y - \sum_i y_i(y) y_i$. Thus y_i, g_i, $1 \leq i \leq m$ is a projective coordinate system for $\mathcal{P}$. The formula $w = \sum_i r_i z_i = (\sum_i r_i y_i) + (\sum_i r_i z_i)$ with $y = \sum_i r_i y_i \in \mathcal{P}$ and $w = \sum_i r_i z_i \in \mathcal{Q}$. means the projection $V \to \mathcal{P}$ is given by $\sum_i r_i w_i \mapsto \sum_i r_i y_i$ and the formula $y = \sum_i g_i(y) w_i$ means the inclusion $\mathcal{P} \to V$ is given by $y \mapsto \sum_i g_i(y) w_i$. Both the projection and the inclusion, we recall, are differential maps.

We use the basis $w_1, \ldots, w_m$ to identify R^m with V. Then there is a matrix $A \in M_m(R)$ such that D_V is identified with D_A. The projection $V \to \mathcal{P}$ then is identified with $\rho : R^m \to \mathcal{P}$ by $\rho([r_1, \ldots, r_m]^T) =$

$\sum_i r_i y_i$ and the inclusion $\mathcal{P} \to V$ is identified with $\sigma : \mathcal{P} \to R^m$ by $\sigma(y) = [g_1(y), \ldots, g_m(y)]^T$. Consider $E = \sigma \circ \rho : R^m \to R^m$. We compute the matrix e of E: let $e_j \in R^m$ be the column whose only non-zero entry is 1 in row j. Then $E(e_i = \sigma(\rho(e_j)) = \sigma(y_j) = [g_i(y_j), \ldots, g_m(y_j)]^T$. This latter is then the jth column of e, so $e = [g_i(y_j)]$. (We saw above that such a mtrix from a projective coordinate system is idempotent).

Since both ρ and σ are differential, so is $E : (R^m, D_A) \to (R^m, D_A)$ which means that $eD_A(x) = D_A(ex)$ for all $x \in R^m$. That is $e(x' + Ax) = (ex)' + Aex$, or $ex' + eAx = e'x + ex' + Aex$ which, to hold for all x, means that $eA = e' + Ae$, so $e' + [A, e] = 0$. We also know by Lemma 4.1 that $e' + [[e, e'], e] = 0$. Thus $[A - [e, e'], e] = 0$. Let $T = A - [e, e']$; then $A = [e, e'] + T$ where $T \in M_m(R)$ commutes with e.

From the projective coordinate system y_i, g_i, $1 \leq i \leq m$ on $\mathcal{P}$, we have that $\mathcal{P} \cong eR^m$ is a differential submodule of $(R^m, D_{[e,e']})$. This differential structure differs from the one obtained from $D_{\mathcal{P}}$, where $\mathcal{P} \cong eR^m$ is a differential submodule of (R^m, D_A), by a linear transformation of eR^m; and we just saw that that linear transformation has the matrix T (which preserves eR^m since T and e commute).

Conversely, if $S \in M_m(R)$ is any matrix that commutes with e, and $B = [e, e'] + S$, then $e : (R^n, D_B) \to (R^m, D_B)$ is differential, since for $x \in R^m$, $D_B(ex) = e'x + ex' + Bex$ while $eD_B(x) = ex' + eBx$ which will be equal for all x provided $e' + [B, e] = 0$. Since $B = [e, e'] + S$, $[B, e] = [[e, e'], e] + [S, e] = [[e, e'], e]$ so $e' + [B, e] = 0$. Thus eR^m is a differential submodule of (R^m, D_B) whose differential structure different from that of eR^m as a differential submodule of $(R^m, D_{[e,e']})$ by the linear transformation of eR^m given by S.

4.3 INDUCED MODULES

Now suppose R is any commutative ring, P is an R module with a projective coordinate system x_i, f_i, $1 \le i \le n$ and R_0 is a subring of R such that $f_i(x_j) \in R_0$ for all $1 \le i, j \le n$. Let $P_0 = \sum_i R_0 x_i \subseteq P$. Note that $f_i(P_0) \subseteq R_0$ for $1 \le i \le n$, so that $g_i = f_i|P_0 : P_0 \to R_0$ makes sense. Then x_i, g_i, $1 \le i \le n$ is a projective coordinate system for P_0. Furthermore, we have maps $p_0 : R_0^n \to P_0$ by $p_0([b_1, \dots, b_n]^T) = \sum_i b_i x_i$ and $q_0 : P_0 \to R_0^n$ by $q_0(y) = [g_1(y), \dots, g_n(y)]^T$ analogous to $p : R^n \to P$ and $q : P \to R^n$ above, where p_0 is surjective, q_0 is injective, and $p_0 \circ q_0$ is the identity on P_0.

Consider the map $h : R \otimes_{R_0} P_0 \to P$ by $\sum_k r_k \otimes y_k \mapsto \sum_k r_k y_k$. It is surjective since $x_i \in P_0$, $1 \le i \le n$. We are going to show it is an isomorphism. We have a commutative diagram

$$
\begin{array}{ccccc}
R \otimes_{R_0} R_0^n & \xrightarrow{1 \otimes p_0} & R \otimes_{R_0} P_0 & \xrightarrow{1 \otimes q_0} & R \otimes_{R_0} R_0^n \\
{\scriptstyle l}\big\downarrow & & {\scriptstyle h}\big\downarrow & & {\scriptstyle l}\big\downarrow \\
R^n & \xrightarrow[p]{} & P & \xrightarrow[q]{} & R^n
\end{array}
$$

where $l : R \otimes_{R_0} R_0 \to R^n$ is the isomorphism $\sum_k r_k \otimes w_k \mapsto \sum_k r_k w_k$. Note that $1 \otimes p_0$ is surjective since p_0 is and $1 \otimes q_0$ is injective since $(1 \otimes p_0) \circ (1 \otimes q_0) = 1 \otimes 1$ is the identity on $R \otimes_{R_0} P_0$.

Suppose $z = \sum_k r_k \otimes y_k$ is in the kernel of h. Then $qh(z) = 0$, hence so does $l(1 \otimes q_0(z))$, and since q_0 is injective and l is an isomorphism this means $z = 0$. Thus h is indeed an isomorphism.

When R_0 is a subring of the ring R, and M_0 is an R_0 module, it is customary to call the R module $M \otimes_{R_0} M_0$ an *induced* module, and we follow this convention, although as Leonard Scott has pointed out, the proper terminology may well be "produced" module, with induced module reserved for the R module $\mathrm{Hom}_{R_0}(R, M_0)$.

We retain these notations and now resume our assumption that R is a differential ring and further suppose that R_0 is a differential subring. It is clear from the formulas of Theorem 4.1 that the differential structure on P_0 from the matrix $e = [f_i(x_j)]$, which presents P_0 as a differential direct summand of $(R_0^n, D_{[e,e']})$ as well as presenting P as a differential direct summand of $(R^n, D_{[e,e']})$, is such that the isomorphism $R \otimes_{R_0} P_0 \to P$ is an isomorphism of differential modules.

An important special case occurs when R_0 is the ring R^D of constants of R. In this case the assumption that $e = [f_i(x_j)] \in M_n(R^D)$ is the statement that $e' = 0$. Now $e' = 0$ certainly implies $[e, e'] = 0$, but its also ture that $[e, e'] = 0$ implies $e' = 0$. For if e and e' commute, then since $e = e^2$, $e' = ee' + e'e = ee' + ee'$; multiply this equation by e to get $ee' = ee' + ee'$ so $ee' = 0$ and therefore $e' = ee' + ee' = 0$.

When P is a finitely generated projective module as in Theorem 4.1 and, in the notation of the theorem, if $[e, e'] = 0$, then P has a differential structure such that it is a direct summand of (R^n, D_0). Moreover, since $[e, e'] = 0$, $e' = 0$ so $e \in M_n(R^D)$. The ambient differential modules $((R^D)^n, D_0)$ and (R^n, D_0) of which P_0 and P are (isomorphic to) differential direct summands are, respectively, R^n with coordinatewise differentiation and $(R^D)^n$ with zero derivation. Any direct summand of $(R^D)^n$ is automatically a differential direct summand, with the zero derivation, and moreover, the zero derivation is the only possibility for a derivation of any differential direct summand of $(R^D)^n$, so the differential structure on P_0 is unique. Direct summands of R^n, however, may not be differential direct summands. Of course, since they are projective, they do carry some differential structure; however, it may not be induced from a (projective) R^D module by tensoring up. We now clarify this relationship.

Proposition 4.3 *Let P be a differential R module, finitely generated and projective as an R module. Then P is induced from R^D as a differential module (that is, there is a finitely generated projective R^D module P_0 such that $P \cong R \otimes_{R^D} P_0$ as differential modules) if and only if P is a differential direct summand of (R^n, D_0) for some n. If it is a differential direct summand then P^D is a finitely generated projective R^D module and $r \otimes x \mapsto rx$ s a differential isomorphism $P \cong R \otimes_{R^D} P^D$.*

Proof. Suppose that $P \cong R \otimes_{R^D} P_0$ as differential modules, where P_0 is a finitely generated projective R^D module. Then there is an R^D module Q_0 and an n such that $(R^D)^n \cong P_0 \oplus Q_0$. We tensor this isomorphism with R over R^D (which distributes over direct sum to obtain $R \otimes_{R^D} (R^D)^n \cong (R \otimes_{R^D} P_0) \oplus (R \otimes_{R^D} Q_0)$. Since $R \otimes_{R_0} (R^D)^n \cong R^n$ with coordinate-wise derivation, we have P displayed as a differential direct summand of (R^n, D_0).

Conversely, let P be a differential direct summand of R^n with coordinate-wise derivation, and let Q be a complementary differential summand: $R^n \cong P \oplus Q$. Apply $(\circ)^D$ to both sides: we have $(R^D)^n \cong (R^n)^D \cong P^D \oplus Q^D$. This shows that P^D and Q^D are finitely generated projective R^D modules. Now tensor this last isomorphism with R over R^D. We have a commutative diagram of differential homomorphisms

$$\begin{array}{ccc} R \otimes_{R^D} (R^D)^n & \longrightarrow & (R \otimes_{R^D} P^D) \oplus (R \otimes_{R^D} Q^D) \\ \downarrow & & \downarrow \\ R^n & \longrightarrow & P \oplus Q \end{array}$$

where the horizontal maps are isomorphisms and the vertical maps are $r \otimes x \mapsto rx$. Since the left vertical map is an isomorphism, so is the right, which means in particular that $P \cong R \otimes_{R^D} P^D$ is induced from a finitely generated projective R^D module.

We consider the implications of Proposition 4.3 when the module P is free as an R module; that is, when $P = (R^n, D_A)$ for some n and some $A \in M_n(R)$. If P is induced, the projective R^D modules $P_0 = P^D$ has the property that $R \otimes_{R^D} P_0$ is free of rank n as an R module. One way this could happen is if P_0 itself is free of rank n as an R^D module. Assume that this is the case. Then, as a differential module $R \otimes_{R^D} P_0 \cong (R^n, D_0)$. Thus, provided that if (R^n, D_A) is induced, $(R^n, D_A)^D$ is a free R^D module, we conclude that (R^n, D_A) is induced (that is, (R^n, D_A) is a differential direct summand of (R^m, D_0) for some m) if and only if $(R^n, D_A) \cong (R^n, D_0)$. One way to ensure that P_0 is a free R^D module is to consider the case that all finitely generated projective R^D modules are free modules. Then, as we noted, induced differential projective R modules are not only free, but have a basis of constants. We state this as a corollary to Proposition 4.3:

Corollary 4.2 *Assume all finitely generated projective R^D modules are free. Then a differential finitely generated projective R module P is a differential direct summand of (R^n, D_0) for some n if and only if $P \cong (R^m, D_0)$ for some m.*

Proof. If P is a differential direct summand of (R^n, D_0) then by Proposition 4.3 $P \cong R \otimes_{R^D} P_0$ for some finitely generated projective, hence free, R^D module P_0, say $P_0 \cong (R^D)^m$. Then $P \cong R \otimes_{R^D} (R^D)^m \cong (R^m, D_0)$. The converse is trivial with $m = n$.

Continue to assume that finitely generated projective R^D modules are free. Let $A \in M_n(R)$ be such that (R^n, D_A) is *not* differentially isomorphic to (R^n, D_0). Then by Corollary 4.2 (R^n, D_A) can not be a differential direct summand of (R^m, D_0) for any m. In particular, there can't be any m, or p and $B \in M_n(R)$, such that differentially $(R^n, D_A) \oplus (R^p, D_B) \cong (R^m, D_0)$. In this notation, let $C \in M_m(R)$ be the matrix

$$\begin{bmatrix} A & 0 \\ 0 & B \end{bmatrix}$$

We have that (R^m, D_C) and (R^m, D_0) are not isomorphic. We recall that this means that there is no invertible matrix $S_m \in M_m(R)$ such that $S_m^{-1}S_m' = C$. In other words, still under the assumption that finitely generated projective R^D modules are free, the fact that there is no $S_n \in GL_n(R)$ such that $S_n^{-1}S_n' = A$ means that for every $m = n+p$ there is no $S_m \in GL_m(R)$ such that

$$S_m^{-1}S_m = \begin{bmatrix} A & 0 \\ 0 & B \end{bmatrix}.$$

In the projective module theory of commutative rings, a finitely generated projective module P is *stably free* if there is a free module R^m such that $P \oplus R^m$ is free ($P \oplus R^m \cong R^m$ for suitable n). There are examples of non-free stably free projective modules. If there are none, then the ring is said to have *cancellation* (of free modules in direct sums; note that it is sufficient to have cancellation just for the case $m = 1$ to have it in general). Let us temporarily call the differential modules of the form (R^n, D_0) constant basis differential modules. Again temporarily, let us say that a differential finitely generated projective module P is stably a constant basis module if there is a differential finitely generated module Q such that $P \oplus Q$ is a constant basis module. Let us continue to assume that finitely generated projective R^D modules are free, and let us further assume that finitely generated R modules are free. Then if P is a stably constant basis differential R module, it is a constant basis R module. For the assumption on R means that P has the form (R^n, D_A), and then our previous discussion shows why P being "stably constant basis" implies being "constant basis". In analogy with the non-differential case, we say R has *(differential) cancellation* (of the differential module R in direct sums).

4.4 TRIVIALIZING EXTENSIONS

We have the functor $M \mapsto R \otimes_{R^D} M$ from the category of R^D modules to the category of differential R modules (or just to the category of R modules, although we will look at this latter as the composition of the tensor up functor with the forgetful functor from differential R modules to R modules). We also have the functor $M \mapsto M^D$ from the category of differential R modules to the category of R^D modules. We can also consider the compositions $M \mapsto (R \otimes_{R^D} M)^D$ and $M \mapsto R \otimes_{R^D} M^D$. These are natural transformations from the identity, and to the identity, respectively, for these compositions, namely $M \to (R \otimes_{R^D} M)^D$ by $m \mapsto 1 \otimes m$ and $R \otimes_{R^D} M^D \to M$ by $r \otimes m \mapsto rm$. Neither need be a natural isomorphism. Since they are both additive, and the first is an isomorphism for $M = R^D$ we know the first is an isomorphism for finitely generated projective R^D modules. However, the second is not, even if we restrict the domain to differential finitely generated projective modules: it is an isomorphism for $M = R$, and therefore for (R^n, D_0) and therefore for direct summands of the latter, but in general the map is neither injective nor surjective.

Philosophically, differential module theory should not provide any insight into the category of R^D modules, or its subset of finitely generated projective R^D modules. What we hope from differential module theory is to provide insight into the category of differential modules, at least, for the present case, those finitely generated and projective as modules, *modulo* the contributions from R^D modules. We know that these differential finitely generated projective modules are differential direct summands of differential modules free of finite rank as modules, and we know at least how to see these latter in terms of the matrices which describe their derivations. When these matrices are equivalent to the zero matrix, then Proposition 4.3 tells us the differential summands are induced. We have seen in previous chapters, how when presented with a differential module which is free of finite rank as a module, how to extend scalars to make the matrix defining its derivation equivalent to the zero matrix. In the present context, this can be used to see how differential finitely generated projective modules can be seen as induced modules over the constants of the extended ring of scalars.

To begin, let (P, D_P) be a differential R module, with P finitely generated and projective as an R module. There exist many finitely generated projective R modules Q such that $P \oplus Q$ is R free. Pick one, and also choose a differential structure on D_Q on it. Then $(P, D_P) \oplus (Q, D_Q)$ is a differential R module, free of finite rank, say n, as an R module. Thus, as differential modules, $P \oplus Q \cong (R^n, D_A)$ for some $A \in M_n(R)$. Let y_{ij}, $1 \leq i, j \leq n$ be n^2 indeterminates over R and let $S_0 = R[\{y_ij | 1 \leq i, j \leq n\}]$

be the polynomial ring they generate over R. Let $Y = [y_{ij}] \in M_n(S_0)$. Define a derivation on S_0 extending D_R such that on $M_n(S_0)$ we have $Y' = YA$. Extend this derivation to $S = S_0[\det(Y)^{-1}]$. Then in $M_n(S)$ we have that Y is invertible and that $Y^{-1}Y' = A$. Thus $S \otimes_R (R^n, D_A) \cong (S^n, D_A) \cong (S^n, D_0)$ where the final differential isomorphism is given by multiplication by Y. Since P is a differential direct summand of (R^n, D_A), $S \otimes_R P$ is (differentially isomorphic to) a differential direct summand of $S \otimes_R (R^n, D_A) \cong (S^n, D_0)$. Thus by Proposition 4.3 $S \otimes_R P)^D$ is a finitely generated projective S^D module and $S \otimes_R P \cong S \otimes_{S^D} (S \otimes_R P)^D$.

Note that S, being a localization of a polynomial ring over R is flat as an R module. In fact, since S is an augmented R algebra (there is a ring homomorphism $S \to R$ determined by $Y \mapsto I_n$) in which R appears as an R module direct summand, S is faithfully flat. And S is finitely generated as an R algebra, by the n^2+1 elements y_{ij}, $1 \leq i,j \leq n$ and $\det(Y)^{-1}$. We pause the discussion to note that we have thus established:

Proposition 4.4 *Let P be a differential finitely generated projective R module. Then there is a finitely generated faithfully flat augmented R algebra S, which is a differential extension of R such that $(S \otimes_R P)^D$ is a finitely generated projective S^D module and $S \otimes_R P \cong S \otimes_{S^D} (S \otimes_R P)^D$.*

To resume the discussion with the same notation, suppose that P is free as an R module. Then we can take $Q = 0$ so that $P \cong (R^n, D_A)$. Then $S \otimes_R P \cong (S^n, D_0)$ so $(S \otimes_R P)^D \cong (S^D)^n$; Proposition 4.4 is, in this case, just the construction we considered in Chapter 1.

The ring S above, as an R algebra, is a localization of a polynomial ring over R. Thus, if R is a localization of a polynomial ring over a field,

S is a localization of a polynomial ring over the same field. If that field is algebraically closed of characteristic zero (for example, $\mathbb{C}$), then R and S are coordinate rings of open subsets of affine space over that field. Moreover, all finitely generated projective modules over R and S are therefore free so the differential modules under consideration are all of the form (R^n, D_A) or (S^m, D_B). Nonetheless, as the following example shows, complications can arise, even in low dimensions.

4.5 AN EXTENDED EXAMPLE

It will be more convenient in this example to consider the row, rather than column, free modules, what we have previously denoted $(R^{(n)}, {}_A D)$.

Let R be the localized polynomial ring $\mathbb{C}[a,b][h^{-1}]$ where $h = a^2 + b$ and $D(a) = a$, $D(b) = b$, and $D(\alpha) = 0$ for $\alpha \in \mathbb{C}$. (The reason for the inversion of h will appear below.) The polynomial ring is, of course, a unique factorization domain; since h is linear in b, it is irreducible. Any element of R can be written as $h^k p$ where k is an integer and p is a polynomial relatively prime to h. We can also write an element $x \in R$ as a fraction $x = \frac{f}{g}$ where $g = h^k$, $k \geq 0$, but with k minimal with respect to such a representation of x as a fraction.

The units of R are the elements αh^k where $\alpha \in \mathbb{C} - \{0\}$ and $k \in \mathbb{Z}$. Such elements are certainly units. To see that all units have this form, suppose $x = h^k p \in R$ with p and h relatively prime, has an inverse $h^l q$ with h and q relatively prime. Then $h^k p h^l q = 1$ so $pq = h^{-k-l}$. Since pq is a polynomial, so is h^{-k-l} and so $-k-l \geq 0$. If $k-l > 0$ then both sides have positive degree and since h divides the right-hand side it divides the left, namely $h|pq$ which is impossible since h is relatively prime to both p and q. Thus $-k-l=0$ which means pq, and hence p and q are inverse elements of $\mathbb{C}$. In particular, $x=\alpha h^k$ with $\alpha \in \mathbb{C}$ as claimed.

A simple calculation shows that $D(a^ib^j)=(i+j)a^ib^j$. A polynomial $p \in \mathbb{C}[a,b]$ can be written uniquely in the form $\sum_{i\geq 0} p_i$ where p_i is homogeneous of degree i. Then $D(p)=\sum_i ip_i$. For example, $h=\sum h_i$ where $h_0=0$, $h_1=b$, $h_2=a^2$ and $h_i=0$ for $i>2$. And $D(h)=b+2a^2$. Note that h does not divide $D(h)$.

Suppose $x \in R$ satisfies $D(x)=x$. We write $x=h^kq$ where h does not divide q. Suppose $k \neq 0$.Then $D(x)=kh^{k-1}D(h)q+h^kD(q)$. Using that $D(x)=x$ and $D(h)=2a^2+b=h+a^2$ we have $h^kq=kh^{k-1}(h+a^2)q+h^kD(q)$. Multiplying through by h^{1-k} and rearranging terms then gives $h(q-k-D(q))=ka^2q$. Since h is a factor on the left, it is a factor of ka^2q, contradicting the facts that h is of positive degree and relatively prime to a^2 and, by assumption q. Thus $k=0$ and $x=q$ is a polynomial. Write $q=\sum_i q_i$. Then $D(q)=q$ and $D(q)=\sum_i iq_i$ implies that $q=q_1$, or $x \in \mathbb{C}a+\mathbb{C}b$.

Now consider the module $M=(R^{(1)},{}_AD)$ where $A=-I_1=[-1]$. Explicitly, $M=R^{(1)}$ with derivation $D_M=m'-m$, or $D_M((x))=(x'-x)$ for $x \in R$.

Note that $D_M((a))=0$ and $D_M((b))=0$ so (a) and (b) are in M^D. Also $a(a)+(b)=(h)$ so $(1)=\frac{a}{h}(a)+\frac{1}{h}(b)$ so that (a) and (b) span M; thus $M=RM^D$, hence $R \otimes_{R^D} M^D \to M$ is surjective.

Next we compute R^D, which turns out to be just $\mathbb{C}$:

If $\frac{p}{q}$, where p and q are relatively prime polynomials where $q=h^k$ is a non-negative power of h, is constant, then $qD(p)-pD(q)=0$.

So $q|D(q)$. If $k>0$, we have h^k as a factor of $kh^{k-1}D(h)$ so $h|D(h)$; as noted, this is not possible.

So $q = 1$ and p is a constant. Write $p = \sum_{i=0}^{n} p_i$, where p_i is the term in p of total degree i.

Then $D(p) = \sum^n ip_i$ so $D(p) = 0$ implies that $p = p_0 \in \mathbb{C}$. Thus $R^D = \mathbb{C}$ as claimed.

We already know that M^D is a complex vector space of dimension 2, which makes M^D a free R^D module of rank 2. So $R \otimes_{R^D} M^D$ is a free R module of rank 2. We also know that $R \otimes_{R^D} M^D \to M$ is surjective. However, M, of course, is a free R module of rank 1. So M is not induced from M^D. By Proposition 4.3 we know that M is not a differential direct summand of $(R^{(n)},_0 D)$ for any n. This, of course, includes that M is not isomorphic to $(R^{(1)},_0 D)$.

We also know that M is generated as an R module by the two constants (a) and (b). We can express this latter fact by noting that $T : R^{(2)} \to M$ by $(x, y) \mapsto x(a) + y(b) = (xa + yb)$ is a differential surjection.

The matrix of T is

$$\begin{bmatrix} a \\ b \end{bmatrix}$$

Suppose there is a right inverse S with matrix

$$B = \begin{bmatrix} x_{11} & x_{12} \end{bmatrix}$$

$S : M \to R^{(2)}$ differential means

$$(mB)' = (m' - m)B$$

$$m'B + mB' = m'B - mB$$

$$B' = -B$$

$$x'_{ij} = -x_{ij}$$

We calculate the elements in R which satisfy this latter, that is, whose derivatives are their negatives.

Suppose $x \in R$ and $x' = -x$. Write $x = \frac{p}{q}$, where p and q are relatively prime polynomials and $q = h^k$ is a power of h

$$\frac{p}{q} = \frac{pD(q) - qD(p)}{q^2}$$

$$\begin{aligned} pq^2 &= qpD(q) - q^2D(p) \\ q(p + D(p)) &= pD(q) \\ h^k(p + D(p)) &= kh^{k-1}pD(h) \\ h(p + D(p)) &= kpD(h) \end{aligned}$$

If $k \neq 0$ this last equation implies that $h|D(h)$, which is not true. Thus $k = 0$ so $x = p$ is a polynomial $p = \sum_{i=0}^{n} p_i$, where p_i is the term in p of total degree i and $D(p) = \sum^{n} ip_i$. Thus $ip_i = -p_i$ for all i, which implies $p = 0$. So $x_{ij} = 0$ for all i, and in particular $ST \neq I_1$. Thus there is no differential right inverse to T.

On the other hand, since M is projective (indeed free) as an R module, T must have an R module right inverse, which it does:

$$\begin{bmatrix} \frac{a}{h} & \frac{1}{h} \end{bmatrix} \cdot \begin{bmatrix} a \\ b \end{bmatrix} = \begin{bmatrix} 1 \end{bmatrix}.$$

We note that R, being a polynomial ring with an irreducible element, has all finitely generated projective modules free, as does the field $\mathbb{C} = R^D$. Thus all differential finitely generated projective are of the form $(R^{(n)},_C D)$. And Corollary 4.2 applies to all differential projective R modules.

R is the coordinate ring of an affine open subset of the complex affine plane, namely the plane minus the closed curve $a^2+b=0$, which is a parabola through the origin.

In this example, M is generated by constants but doesn't have a basis of constants. That means that M must be generated by at least two constants, and in fact, it is generated by two, (a) and (b). Now it happens that (a) and (b) are a $R^D = \mathbb{C}$ basis of M^D; since $\mathbb{C} \subset R$ any $\mathbb{C}$ basis of M^D will also generate M. This is a general principal: for any R with R^D a field and any M if M^D is finite dimensional over R^D and $M = RM^D$ then any R^D basis of M^D generate M. It is possible that a full basis of M^D is not required. Here is an example, closely related to the one just considered. Because of the close relation, some of the calculations above will not be repeated.

Let $R = \mathbb{C}[a,b,c][h^{-1}]$ where $h = a^2b^2 + a^2c + b^2c$; note that h is linear in c and hence irreducible in the polynomial ring $\mathbb{C}[a,b,c]$'. Make R differential by defining $D(a) = a$, $D(b) = b$, $D(c) = c$, and $D(\alpha) = 0$ for $\alpha \in C$.

A simple calculation shows that $D(a^ib^jc^k) = (i+j+k)a^ib^jc^k$. A polynomial $p \in \mathbb{C}[a,b,c]$ can be written in the form $p = \sum_i p_i$ where p_i is homogeneous of total degree i. Then $D(p) = \sum_i ip_i$. For example, $D(h) = 4a^2b^2 + 3(a^2c + b^2c)$. For later use we note that $D(h) = 3h + a^2b^2$.

Suppose $y \in R^D$. Write $y = h^kz$ where k is an integer and $z \in \mathbb{C}[a,b,c]$ is relatively prime to h. First suppose $k \neq 0$. Then $0 = D(y) = kh^{k-1}z + h^kD(z)$. Multiply through by h^{1-k} and rearrange terms to get $kz = -hD(z)$. This implies that h divides z, contrary to assumption. Thus $k = 0$ and $y = z$. Now write $z = \sum_i z_i$ where z_i is homogeneous of degree

i. Then $D(y) = D(z) = 0$ implies that $\sum_i iz_i = 0$ so that $z_i = 0$ for $i > 0$. Thus $z \in \mathbb{C}$, showing that $R^D = \mathbb{C}$.

Suppose $x \in R$ and $D(x) = x$. Write $x = h^k q$ with k an integer and $q \in \mathbb{C}[a,b,c]$ and h relatively prime to q. Suppose $k \neq 0$. Using that $D(x) = x$ and $D(h) = 3h + a^2b^2$, we have $kh^{k-1}3hq + kh^{k-1}a^2b^2q + h^k D(q) = h^k q$. Multiplying by h^{1-k} and rearranging terms gives $ka^2b^2q = h(q - 3kq - D(q)$ which contradicts the fact that h noes not divide q. Thus $k = 0$ and $x = q$. Now we write q as $\sum_i q_i$ where q_i is homogeneous of degree i. Then $D(q) = \sum_i iq_i$ so $D(q) = q$ implies $q_i = 0$ for $i \neq 1$. We conclude that x is a homogeneous polynomial of degree one. Thus for $x \in R$, $D(x) = x$ if and only if $x \in \mathbb{C}a + \mathbb{C}b + \mathbb{C}c$.

Now consider the module $M = (R^{(2)}, {}_A D)$ where $A = -I_2$. Explicitly, $M = R^{(2)}$ with $D_M(m) = m' - m$, or $D_M((x,y)) = (x' - x, y' - y)$.

Note that $D_M((a,0)) = (0,0)$, $D_M((0,b)) = (0,0)$, and $D_M((c,c)) = (0,0)$, so $(a,0)$, $(0,b)$, and (c,c) are in M^D.

We have that $a(a,0) + (c,c) = (a^2 + c, c)$ and $b(0,b) + (c,c) = (c, b^2 + c)$ are the rows of a 2×2 matrix whose determinant is h and hence the determinant is a unit of R, and thus $(a^2 + c, c)$ and $(c, b^2 + c)$ t form a basis of M. Since both basis elements are linear combinations of them, the constants $(a,0)$, $(0,b)$, and (c,c) span M. That is, $M = RM^D$ and $R \otimes_{R^D} M^D \to M$ is surjective.

Suppose $(x,y) \in M^D$: $(0,0) = D_M((x,y)) = (x' - x, y' - y)$ so $D(x) = x$ and $D(y) = y$. Conversely, if $x' = x$ and $y' = y$ then $(x,y) \in M^D$. Thus $M^D = \{(x,y) | x' = x \text{ and } y' = y\}$. By our calculation above, this means $M^D = \{(x,y) | x, y \in \mathbb{C}a + \mathbb{C}b + \mathbb{C}c\}$. Note that this means M^D is a 9 dimensional complex vector space with basis $(a,a), (a,b), (a,c), (b,a), (b,b), (b,c), (c,a), (c,b), (c,c)$; since $R^D = \mathbb{C}$ hence

$R \otimes_{R^D} M^D$ is a free R module of rank 9, which we view as being a free R module on the same basis.

Suppose (x,y) and (z,w) are an R basis of M consisting of constants. That is, they span $R^{(2)}$ and satisfy $D(x) = x$, $D(y) = y$, $D(z) = z$, and $D(w) = w$.

The determinant of the matrix with rows (x,y) and (z,w) is $g = xw - yz$ and $D(g) = D(x)w + xD(w) - D(y)z - yD(z) = xw + xw - yz - yz = 2g$; since the matrix is invertible g is a unit. Since $D(g) = 2g$, $D(g) \neq 0$, so $g \notin \mathbb{C}$.

R is the localization of a complex polynomial ring at the irreducible polynomial h so as above its units are complex multiples of powers of h.

So $g = ch^k$, for some $c \in \mathbb{C}$, $c \neq 0$, and some $k \in \mathbb{Z}$, $k \neq 0$. Then $2g = 2ch^k$ while $D(g) = ckh^{k-1}D(h)$ so $2ch^k = kh^{k-1}D(h)$. Multiply this last equation by h^{1-k}, we obtain $2ch = kD(h)$ so $h|D(h)$ in the polynomial ring $\mathbb{C}[a,b,c]$. However, $D(h) = 4a^2b^2 + 3b^2c + 3a^2c$ is not a polynomial multiple of h. This contradiction shows that M does not have a basis of constants, despite the fact that it is spanned by constants (constantly generated).

We already know that $R^D = \mathbb{C}$, that $M = RM^D$, and that M^D is a complex vector space of dimension 9. So any complex basis of M^D will generate M as an R module; the three generators $(a,0)$, $(0,b)$, and (c,c) are part of the basis of M^D identified above, but only part.

We can express the fact that that M is generated as an R module by the three constants $(a,0),(0,b)$, and (c,c). by noting that $T : R^{(3)} \to M$ by $(x,y,x) \mapsto x(a,0) + y(0,b) + z(c,c) = (xa + zc, yb + zc)$ is a differential surjection when $R^{(3)}$ has coordinate wise derivation. We note that if T were differentially split, then M would be a differential direct summand of $R^{(3)}$ and hence have a basis of constants, which we know is not true. Explicitly, T is given by the matrix

$$B = \begin{bmatrix} a & 0 \\ 0 & b \\ c & c \end{bmatrix}$$

Suppose there were a differential map $S : M \to R^{(3)}$ such that TS is the identity on $R^{(2)}$. S would be given by a 2×3 matrix

$$C = \begin{bmatrix} x_{11} & x_{12} & x_{13} \\ x_{21} & x_{22} & x_{23} \end{bmatrix}$$

For $S : (R^{(2)},_{-I_2} D) \to (R^{(3)},_0 D)$ to be differential, C has to satisfy

$${}_0D((x,y))C) = ({}_{-I}D((x,y))C)$$

for all $x, y \in R$. The left-hand side of this equation is ${}_0D((x,y))C + (x,y)C'$, or $(x',y')C + (x,y)C'$ while the right-hand side is $(x'-x, y'-y)C$ or $(x',y')C - (x,y)C$. If we cancel the common term $(x',y')C$ from both sides, we are left with the equation

$$(x,y)C' = -(x,y)C.$$

Since this holds for all $x, y \in R$ we conclude that $C' = -C$ or, in terms of matrix entries, $x'_{ij} = -x_{ij}$ for all i, j. But can the derivative of an element x of R be its own negative, leaving aside the trivial case $x = 0$? Obviously, x cannot be a constant. Suppose $x = h^k y$ with y a polynomial relatively prime to h and $k \neq 0$. Then $D(x) = -x$ implies that $kh^{k-1}D(h)y + h^k D(y) = -h^k y$. As usual, multiplication by h^{1-k} and rearrangement gives $kD(h)y = -h(D(y) + y)$, and as $D(h)$ and h

are relatively prime, this would lead to the contradiction that h divides y. Therefore $k = 0$ and $x = y$. Then we write y as a sum of homogeneous polynomials $y = \sum_i y_i$. Then $D(y) = \sum_i iy_i$ and $D(y) = -y$ together say that $y_i = -iy_i$ for all i. This implies that $y_0 = 0$ and that $y_i = 0$ for $i > 1$. And $y_1 = -y_1$ also implies that $y_1 = 0$. We conclude that $y = 0$ is the only element of R satisfying $D(y) = -y$. Thus, for C, $x_{ij} = 0$ for all i, j, and $C = 0$ is the only possible matrix that gives a differential homomorphism $S : (R^{(2)},_{-I} D) \to (R^{(3)},_0 D)$, but the zero map S can't satisfy $TS = I$.

4.6 RANK ONE DIFFERENTIAL PROJECTIVE MODULES

Our first example in the preceding section was of a differential module whose underlying module was free of rank one. For nay differential ring R, we can consider differential modules which are projective of rank one as R modules. Any such module is finitely generated as an R module, and hence carries differential structures. If I is a rank one projective R module then $R \to \mathrm{Hom}_R(I, I)$ by $r \mapsto L_r$ where $L_r(x) = rx$ is an isomorphism of rings. Suppose D_I is a differential structure on I making I a differential R module. Then $D(L_r)(x) = D_I(L_r(x)) - L_r(D_I(x)) = D_I(rx) - rD_I(x) = D_R(r)x + rD_I(x) - rD_I(x) = D_R(r)x = L_{D(r)}x$. Thus $D(L_r) = L_{D(r)}$, so that $R \to \mathrm{Hom}_R(I, I)$ is differential.

Since the rank one projective R module I is finitely generated, $I^* \otimes_R I \to \mathrm{Hom}_R(I, I)$ by $(f \otimes x)(y) = f(y)x$ is an isomorphism which we have previously seen to be differential when I is. Thus for any rank one projective R module and and differential structure on I, we have $I^* \otimes I \cong (R^{(1)},_0 D)$ as differential modules.

Any differential rank one projective R module which is free as an R module is of the form $(R^{(1)},_A D)$ where A is a matrix with a single entry, say $A = [a]$. For convenience in this section we will simplfy notation and write (R,a) for $(R^{(1)},_{[a]} D)$. Thus for $x \in (R,a)$, $D(x) = x' - ax$. Consider $(R,a) \otimes (R,b)$: $D(x \otimes y) = D(x) \otimes y + x \otimes D(y) = x' \otimes y + ax \otimes y + x \otimes y' + x \otimes by = (x \otimes y)' + (a+b)(x \otimes y)$. Thus with the isomorphism $R \otimes_R R \to R$ by $x \otimes y \mapsto xy$, we have differentially $(R,a) \otimes (R,b) \cong (R, a+b)$. Now technically $(R^{(1)},_A D)^* \cong (R^1, D_{-A^T})$. But since we are dealing with 1 vectors, we can also write R for R^1, and since we are dealing with 1×1 matrices, which are symmetric we have $A = A^T$ so we can write $(R,a)^* \cong (R,-a)$ naturally. If (R,a) is differentially isomorphic to (R,b) then $(R,a) \otimes (R,-b)$ is differentially isomorphic to $(R,b) \otimes (R,-b)$. Applying the above formula for the tensor product then gives $(R, a-b)$ differentially isomorphic to $(R,0)$. Conversely, if $(R,a-b)$ is differentially isomorphic to $(R,0)$ then tensoring both with (R,b) and applying the tensor formula gives (R,a) differentially isomorphic to (R,b). Thus to determine differential isomorphisms between R free differential rank one projective R modules it is enough to determine when (R,a) is isomorphic to $(R,0)$. As we have established previously, this happens if and only if there is an invertible 1×1 matrix, namely an invertible element $s \in R$,with $s's^{-1} = a$.

The isomorphism class of a rank one projective R module I will be denoted $[I]$. If J is also a rank one projective R module we define $[I] \cdot [J]$ to be $[I \otimes J]$. (This is well defined since isomorphism is an equivalence relation.) Since $[I \otimes J] = [J \otimes I]$ and $[I] \cdot [R] = [I]$ the set of isomorphism classes of rank one projective R modules forms a commutative monoid with identity $[R]$, and since $[I]\cdot[I^*]=[R]$ theu form a group with $[I]^{-1}=[I^*]$, called the Picard group of R and denoted $\mathrm{Pic}(R)$.

We can do the same thing with differential isomorphism classes of differential rank one projective R modules. We use the same notation for isomorphism classes. Thus $[(R,a)]$ denotes the set of differential rank one projective R modules isomorphic to (R,a). We have seen what these are: $[(R,a)]=\{(R,b)|$ there is an invertible $s\in R$ with $b=a+s's^{-1}\}$.To see that $[(R,0)]$ is the identity, we need to check that $(R,0)\otimes I\to I$ by $r\otimes x\mapsto rx$ is a differential isomorphism, which it is since $D(rx)=D(r)x+rD(x)$ while $D(r\otimes x)=D(r)\otimes x+r\otimes D(x)\mapsto D(r)x+rD(x)$. Since differentially $I\otimes I^*$ is isomorphic to $(R,0)$, we again have that the inverse of $[I]$ is $[I^*]$. Thus the differential isomorphism classes of differential rank one projective R modules form a group. This is called the *differential Picard* group of R and is denoted $\mathrm{Pic}_{\mathrm{diff}}(R)$.

There is a group homomorphism $\mathrm{Pic}_{\mathrm{diff}}(R)\to\mathrm{Pic}(R)$ given by forgetting the differential structure. It is surjective since every rank one projective R module carries a differential structure. The kernel is the subgroup of differential R modules whose underlying module is isomorphic to R; that is, the kernel is $\{[(R,a)]|a\in R\}$.

The map $GL_1(R)\to G_a(R)$ by $s\mapsto s's^{-1}$ is usually called the *logarithmic derivative* and denoted dlog; it is a homomorphism. Using it and the observations we have already made, we have the following exact sequence.

Theorem 4.2 *There is an exact sequence*

$$0\to GL_1(R^D)\to GL_1(R)\to G_a(R)\to Pic_{diff}(R)\to Pic(R)\to 0$$

where from the left the nontrivial maps are the inclusion, dlog, $a \mapsto [(R,a)]$*, and the forgetful map.*

4.7 ANOTHER EXAMPLE

Let $R = \mathcal{O}(\mathbb{C})$ be the ring of entire functions on the complex plane in the variable z which is a differential ring with the derivation with $D = \frac{d}{dz}$. By Weierstrass theory, finitely generated ideals of R are principal. As we now see, no proper one of these is a differential ideal.

For suppose $f \neq 0$ is an entire function, and Rf is a differential ideal. Suppose that $f(\alpha) = 0$ for some $\alpha \in \mathbb{C}$, and suppose the order of the zero of f at α is $n > 0$. Then f' has a zero at α of order $n-1$. But if Rf is a differential ideal, $f \in Rf$ implies $f' \in RF$, so $f | f'$ and therefore f' has a zero at α of order at least n. Therefore, f has no zeros, and hence is a unit, so that $Rf = R$. We conclude that R is the only finitely generated differential ideal of R.

However, R has proper differential ideals: again by Weierstrass Theory, there is an entire function g which has a zero at n of order n for every natural number $n \in \mathbb{N}$. For any m the functions $g^{(i)}$, $i = 0, 1, \ldots, m$ have infinitely many common zeros at $m+1, m+2, \ldots$, and so the ideal of R generated by $g, g', \ldots, g^{(m)}$ does not contain 1. So differential ideal I of R generated by g is proper. Then the non-zero differential R module R/I is finitely generated as an R module, but not projective as an R module (R is an integral domain).

Because finitely generated ideals of R are principal, finitely generated submodules of free modules are free. In particular, finitely generated projective R modules are free. So a differential R module M which is finitely generated and projective as R module is R free say of rank n: $M \cong (R^n, D_A)$. for some $n \times n$ matrix A: $D_A(m) = m' + mA$. We know that M has a basis of constants $(M \cong (R^{(n)}, D_0)$ provided there is an invertible $n \times n$ matrix Y with entries in R such that $Y' = AY$. The existence of Y is established analytically as follows.

Both A and Y are matrices of functions on $\mathbb{C}$, so we indicate this by writing $Y = Y(z)$ and $A = A(z)$. Consider the integral equation

$$Y(z) = I_n + \int_0^z A(w)Y(w)dw$$

Then we apply Picard iteration: Let $Y_0(z) = I_n$ and define $Y_m(z)$ iteratively as

$$Y_{m+1}(z) = I_n + \int_0^z A(w)Y_m(w)dw.$$

Each new $Y_m(z) \in M_n(R)$ since integrals of entire functions are entire.

Now Fix $R > 0$ and let

$$M(R) = \sup_{|z| \leq R} ||A(z)||$$

(the choice of norm $||\cdot||$ doesn't matter).

Then for $|z| \leq R$ we have

$$||Y_{m+1}(z) - Y_m(z)|| \leq \frac{M(R)^{m+1}|z|^{m+1}}{(m+1)}.$$

This shows us that the series $\sum_m |Y_{m+1}(z) - Y_m(z)|$ is (uniformly) absolutely convergent on $|z| \leq R$, so that the sequence $Y_0, Y_1, \ldots$ converges absolutely on $|z| \leq R$ to a holomorphic function Y. As R is arbitrary, this implies that there is a holomorphic (matrix) function Y to which the $Y_m(z)$'s converge uniformly everywhere. This implies that

$$Y(z) = \lim_{m \to \infty} (I_n + \int_0^z A(w)Y_m(w)dw) = I_n + \int_0^z A(w)Y(w)dw.$$

Note that this implies that $Y(0) = I_n$.

Then differentiating the matrix equation gives

$$Y'(z) = A(z)Y(z).$$

We apply Liouville's formula to this last equation:

$$\frac{d}{dz}(\det(Y(z)) = \text{trace}(A(z))\det(Y(z)).$$

Since $Y(0) = I_n$, the entire function $\det(Y(z))$ takes the value 1 at the origin and hence is non-zero, so we can talk about its inverse in the field of meromorphic functions (the quotient field of R). In particular, as a meromorphic function $y = \det(Y(z))$ satisfies the first-order linear differential equation

$$(\log(y))' = \text{trace}(A(z))$$

so

$$log(y) = \int_0^z \text{trace}(A(w))dw$$

and

$$\det(Y(z)) = y = e^{\int \text{trace}(A)}.$$

This displays the meromorphic function $\det(Y(z))$ as an exponential, which tells us that it is entire, hence in R, and has no zeros, hence is a unit of R. Thus $Y \in M_n(R)$ is an invertible matrix. We conclude, finally, that $M = (R^{(n)}, _A D)$ has a basis of constants.

(I am grateful to ChatGPT for providing this proof, which means that I am grateful to the uncredited sources who placed the documents that ChatGPT cribbed on the internet.)

CHAPTER 5

K-Theory of Differential Modules

5.1 MONOIDS

Suppose we want to classify[1] or otherwise describe all the projective differential modules over a given differential commutative ring. Projective differential modules form an additive category, but in most cases it is not abelian nor does it have infinite limits. (If it were and it did, we could try to classify projective differential modules by the Mitchell Theorem, by identifying the category as the category of modules over the endomorphism ring of an injective cogenerator.) So the path to describing the category of projective differential modules is not obvious. The same issue arises with classifying projective modules over any commutative ring A. We stick with finitely generated modules. One starts with the set $M(A)$ of isomorphism classes of finitely generated projective

[1]Chapter based on Differential projective modules over differential rings, II, Comm. Alg. 50, 2022, 1421–4133.

DOI: 10.1201/9781003452164-5

modules and notices that these form a commutative monoid using the direct sum: $[P]+[Q] := [P \oplus Q]$ is the operation and $[0]$ is the identity. One could stop at this point and try to describe this monoid, but typically one proceeds to the group $K_0(A)$ generated by $M(A)$ (we will be explicit about what this means shortly). Some information is lost in this process: if P is a finitely generated projective R module, let (P) denote the image of its isomorphism class $[P]$ in $K_0(A)$. If P, Q, and M are finitely generated projective A modules with $P \oplus M \cong Q \oplus M$ then in $K_0(A)$ $(P)+(M) = (Q)+(M)$ so, since $K_0(A)$ is a group, $(P) = (Q)$. Conversely, one can show that if $(P) = (Q)$ there must be an A module M such that $P \oplus M \cong Q \oplus M$. (We will explain this below for monoids in general.)

Now since there is a finitely generated projective A module N such that $M \oplus N$ is free, say $M \oplus N \cong A^n$, then $P \oplus A^n \cong Q \oplus A^n$. In other words, we can assume that M above is free. Thus $K_0(A)$ classifies finitely generated projective A modules up to adding on free modules and isomorphism, which is generally considered sufficient. And working with (abelian) groups is generally straightforward.

We could try to follow the same outline for differential finitely generated projective modules over a differential commutative ring R. The set $M^{\text{diff}}(R)$ of isomorphism classes of differential finitely generated projective modules forms a commutative monoid using the direct sum: if $[\cdot]$ denotes an isomorphism class, then $[P]+[Q] := [P \oplus Q]$ is the operation and $[0]$ is the identity. We then have the group $K_0^{\text{diff}}(R)$ formed from $M^{\text{diff}}(R)$. As before, we use (P) to denote the image of $[P]$ in $K_0^{\text{diff}}(R)$. Again, as before, if $(P) = (Q)$ we can show there is a differential finitely generated projective module M such that $P \oplus M \cong Q \oplus M$. Adding a differential finitely generated projective module N to both sides we agin can assume that the underlying projective module for M is free; that is, $M \cong (R^n, D_A)$ for some n and some $A \in M_n(R)$. This is certainly an analogue to the plain projective situation, but it may be too coarse: instead of classifying up to the addition of free modules, which are determined by rank (and isomorphism) in the plain projective case, we are allowing the addition of free modules with arbitrary derivation (and isomorphism). There is no way to have a finer classification – say up to the addition of free modules with coordinate derivation – and still have a group. But we can do that if we stay with monoids.

First, however, we want to review the construction of the group generated by an abelian monoid. So let M be an abelian monoid, by which we mean M has a commutative associative operation, denoted $+$, and a designated identity, denoted 0. Let $\mathcal{F}(M)$ be a free abelian group on the elements of M: $\mathcal{F}(M) = \sum_{m \in M} \mathbb{Z}z(m)$, where we are using $z(m)$ to denote the basis element of $\mathcal{F}(M)$ corresponding to $m \in M$.

An element $z \in \mathcal{F}(M)$ can be written, uniquely, $z = \sum_{i=1}^{m} a_i z(m_i) - \sum_{j=1}^{m} b_j z(m_j)$ where all the a_i and b_j are positive (one or both of the sums may be empty). We will temporarily call this a canonical difference description.

Let $\mathcal{M}(M) = \{\sum_m a_m z(m) | a_m \geq 0 \text{ for all } m \in M\}$. Note that $\mathcal{M}(M)$ is a submonoid of $\mathcal{F}(M)$ regarded as a monoid. We have a monoid homomorphism $\mu_M : \mathcal{M}(M) \to M$ defined by $\mu(\sum_m a_m z(m)) = \sum_m a_m m$. For $a, b \in M$ consider the following elements of $\mathcal{F}(M)$: $p(a,b) = z(a+b)$, $n(a,b) = z(a) + z(b)$, and $s(a,b) = p(a,b) - n(a,b)$. Note that $\mu(p(a,b)) = a + b = \mu(n(a,b))$. Then let $\mathcal{K}(M)$ be the subgroup of $\mathcal{F}(M)$ generated by $\{s(a,b) | a, b \in M\}$, and let $K(M)$ be the quotient group $\mathcal{F}(M)/\mathcal{K}(M)$. For

$m \in M$ let $x(m) = z(m) + \mathcal{K}(M)$. Note that for $a, b \in M$, since $s(a,b) \in \mathcal{K}(M)$, $s(a,b) = z(a+b) - z(a) - z(b) \in \mathcal{K}(M)$, so $x(a+b) = x(a) + x(b)$. If x is any element of $K(M)$, let $z \in \mathcal{F}(M)$ be such that $x = z + \mathcal{K}(M)$. Let $z = \sum_{i=1}^{m} a_i z(m_i) - \sum_{j=1}^{m} b_j z(m_j)$ where all the a_i and b_j are positive be a canonical difference description of z. Then

$$x = \sum_i a_i x(m_i) - \sum_j b_j x(m_j) = x(\sum_i a_i m_i) - x(\sum_j b_j x(m_j))$$

that is, $x = x(m) - x(n)$ for elements $m, n \in M$.

We have seen that $M \to K(M)$ by $m \mapsto x(m)$ is a monoid homomorphism from M to an abelian group. Moreover, if $f : M \to K$ is any monoid homomorphism from M to an abelian group K, then we have a group homomorphism $F : \mathcal{F}(M) \to K$ determined by $F(z(m)) = f(m)$. Since for $a, b \in M$ $f(a+b) = f(a) + f(b)$, $F(z(a+b)) = f(a+b) = f(a) + f(b) = F(z(a)) + F(z(b)) = F(z(a) + z(b))$. Thus $F(s(a,b)) = 0$, so $F(\mathcal{K}) = \{0\}$ and F induces a group homomorphism $\phi : K(M) \to K$ which satisfies $\phi(x(m)) = f(m)$ for $m \in M$. Since by definition $\mathcal{F}(M)$ is generated by $\{z(m) | m \in M\}$, $K(M)$ is generated by $\{x(m) | m \in M\}$. Thus ϕ is determined by the fact that $\phi(x(m)) = f(m)$ for all $m \in M$, which justifies denoting it ϕ_f. In other words, $K(M)$ has the (universal) property that for any monoid homomorphism $f : M \to K$ where K is a group there is a unique group homomorphism $\phi_f : K(M) \to K$ such that $\phi_f(x(m)) = f(m)$ for all $m \in M$. In this sense we can say that $K(M)$ is the group generated by M; the universal property shows it is unique up to isomorphism.

$K(M)$ is the most general group to which M maps. The most general group which maps into (injectively) M is its subgroup $U(M) = \{m \in M | \exists n \in M \text{ with } m + n = 0\}$ of units.

Let us examine what it means for elements $a, b \in M$ to have $x(a) = x(b)$. Then in $\mathcal{F}(M)$ we have $z(a) = z(b) + \kappa$ where $\kappa \in \mathcal{K}(M)$. We can write $\kappa = \kappa_1 + \kappa_2$ where $\kappa_1 = \sum_i \alpha_i s(c_i, d_i)$ where $\alpha_i > 0$ all i and $\kappa_2 = \sum_j \beta_j s(e_j, f_j)$ where $\beta_j < 0$ all j. Then

$$z(a) + \sum_i \alpha_i n(c_i, d_i) + \sum_j |\beta_j| p(e_j, f_j)$$
$$= z(b) + \sum_j |\beta_j| n(e_j, f_j) + \sum_i \alpha_i n(c_i, d_i).$$

Apply μ to both sides of this equation. Recalling that $\mu(p(a,b)) = \mu(n(a,b)) = a + b$ we then have that

$$a + \sum_i \alpha_i (c_i + d_i) + \sum_j |\beta_j| (e_i + f_i) = b + \sum_j |\beta_j| (e_i + f_i) + \sum_i \alpha_i (c_i + d_i).$$

Let $m = \sum_i \alpha_i (c_i + d_i) + \sum_j |\beta_j| (e_i + f_i)$. Then $a + m = b + m$. Conversely, if there is an element $m \in M$ such that $a + m = b + m$ in M then $x(a) + x(m) = x(a+m) = x(b+m) = x(b) + x(m)$ in $K(M)$, and since $K(M)$ is a group we have $x(a) = x(b)$.

Thus:

Proposition 5.1 *Let a and b be elements of the monoid M. Then $x(a) = x(b)$ in the group $K(M)$ generated by M if and only if there is an element $m \in M$ such that $a + m = b + m$.*

A submonoid N of a monoid M is *cofinal* if given $p \in M$ there is $q \in M$ such that $p + q \in N$. When M is a group, $\{0\}$ is cofinal. When $M = M(A)$, the submonoid $N = \{R^n | n \geq 0\}$ is cofinal. When $M = M^{\text{diff}}(R)$ the submonoid $N = \{(R^n, D_A) | n \geq 0, A \in M_n(R)\}$ is cofinal, but in general $N_0 = \{(R^n, 0) | n \geq 0\}$ is not cofinal.

The notion of cofinal submonoid enables a slight modification of Proposition 5.1:

Corollary 5.1 *Let a and b be elements of the monoid M. Let N be a cofinal submonoid of M. Then $x(a) = x(b)$ in the group $K(M)$ generated by M if and only if there is an element $n \in N$ such that $a+n = b+n$.*

The corollary follows since if, in the notation of the proposition, $a+m = b+m$, there is a $q \in M$ such that $m+q \in N$.

Proposition 5.1 applied to $M = M(A)$ (respectively $M^{\text{diff}}(R)$) explains why, in our earlier informal notation for images of isomorphism classes, $(P) = (Q)$ in $K_0(A)$ (respectively $K_0^{\text{diff}}(R)$) if and only if there is an N such that $P \oplus N \cong Q \oplus N$. As remarked previously, for the case of $K_0(A)$, we may assume $N = A^n$ for some n, and for the case of $K_0^{\text{diff}}(R)$, we may assume $N = (R^n, D_B)$ for some n and some $B \in M_n(R)$. To reiterate, using $K_0^{\text{diff}}(R)$ we may go this fine in the classification of differential modules (classifying up to the addition of modules of the form (R^n, D_A)) but in general not as fine as classifying up to addition of modules of the form $(R^n, 0)$. If we want to use the latter classification, we have to stay in the monoid world.

For the classification of finitely generated projective modules over a commutative ring A, we may very well consider the free modules to be understood, and only try to understand "projectives modulo frees". This is easy to make sense of using the group $K_0(A)$: let $\mathcal{N} \leq K_0(A)$ denote the subgroup consisting of the free modules ($\mathcal{N} = \mathbb{Z}(A)$). Then $\overline{K_0(A)} = K_0(A)/\mathcal{N}$, sometimes called the *reduced* K_0 group of A, classifies isomorphism classes of finitely generated projective A modules modulo those of free modules. Not much is actually lost in this process. Let P be a projective A module and let $|P| = (P) + \mathcal{N}$. Assume that all finitely generated projective A modules have constant rank. Isomorphic finitely generated projective modules have the same rank, so rank is defined on isomorphism classes, and rank is additive on direct sums so rank defines a monoid homomorphism $\rho : MA) \to \mathbb{Z}$. Since $\mathbb{Z}$ is a group, ρ factors through $K_0(A)$ and hence defines a function $\text{rk} : K_0(A) \to \mathbb{Z}$ such that for a projective module P $\text{rk}((P))$ is the rank of P. Consider $\text{Ker}(\text{rk}) \leq K_0(A)$.

Suppose P has rank n. Then $(P) = (P) - (A^n) + (A^n)$ in $K_0(A)$ with $(P) - (A^n) \in \text{Ker}(\text{rk})$. This shows that $K_0(A) = \text{Ker}(\text{rk}) + \mathcal{N}$. Consider $(A^n) \in \mathcal{N}$. Since $\rho((A^n)) = n$, $(A^n) \in \mathcal{N}$ if and only if $n = 0$. Thus $\text{Ker}(\text{rk}) \cap \mathcal{N} = (0)$ so the sum $K_0(A) = \text{Ker}(\text{rk})\mathcal{N}$ is direct (and $\text{rk}|\mathcal{N} : \mathcal{N} \to \mathbb{Z}$ is an isomorphism) which shows that the projection $K_0(A) \to \overline{K_0(A)}$ induces an isomorphism $\text{Ker}(\text{rk}) \to \overline{K_0(A)}$. Thus $K_0(A) \cong \overline{K_0(A)} \oplus \mathbb{Z}$. This means that in regarding free modules as trivial, and only considering projectives modulo frees, nothing is lost.

One further observation about plain projectives: consider $M(A) \to \overline{K_0(A)}$ and suppose $|P| = |Q|$. Then $(P) = (Q) + n(R)$. We can assume $n \geq 0$; if not, move it to the other side of the equation. Thus $(P) = (Q \oplus R^n)$ which means there is an M such that $P \oplus R^m \cong Q \oplus R^{n+m}$. That is, P and Q correspond to the same element of $\overline{K_0(A)}$ if (and only if) there are free modules F_1 and F_2 such that $P \oplus F_1 \cong Q \oplus F_2$.

What happens when we try to mimic these constructions for the differential ring R? We continue to assume that finitely generated projective R modules have constant rank. We define the rank of a differential finitely generated projective R module (P, D_P) to be the rank of its underlying projective module P. Then rank defines a monoid homomorphism from $M^{\text{diff}}(R) \to \mathbb{Z}$ As before, this factors through $K_0^{\text{diff}}(R)$ defining a function $\text{rk} : K_0^{\text{diff}}(R) \to \mathbb{Z}$ such that $\text{rk}((P))$ is the rank of P.

Let $\mathcal{N} = \mathbb{Z}((R, D_0))$ be the subgroup of $K_0^{\text{diff}}(E)$ generated by (R, D_0). We still have that $\text{Ker}(\text{rk}) \cap \mathcal{N} = [0]$. Again, if P has rank n then $(P) = (P) - n(R, D_0) + (R^n, D_0)$ so that $K_0^{\text{diff}}(R) = \text{Ker}(\text{rk}) + \mathcal{N}$ (direct sum), $\text{rk}|\mathcal{N} : \mathcal{N} \to \mathbb{Z}$ is an isomorphism so that $K_0^{\text{diff}}(R) = K_0^{\text{diff}}(R)/\mathcal{N} \oplus \mathbb{Z}$. Again write $|P|$ for $(P) + \mathcal{N}$. Because of the direct sum decomposition, classifying P amounts to understanding $|P|$. So what does $|P| = |Q|$ mean? Initially, that $(P) = (Q) + n((R, D_0))$ with $n \geq 0$. So $(P) = (Q + (R^n, D_0)$, which in turn implies $P \oplus (R^m, D_A) \cong Q \oplus (R^{n+m}, D_B)$ where $B = 0 \oplus A \in M_{n+m}(A)$. Now B and A are not unrelated, but the modules (R^m, D_A) and (R^{n+m}, D_B) are not linked in the same way that the free modules F_1 and F_2 were in the plain projective case. It would be nice, for example, if $|P| = |Q|$ implied that $P \oplus (R^p, D_0) \cong Q \oplus (R^q, D_))$ for some p, q, which would be a close analogue of the plain projective case. However, as we have seen, the fact that we are dealing with the *group* $K_0^{\text{diff}}(R)$ means that the best we can do with $(P) = (Q)$ is up to the addition of an (R^n, D_A) where A need not be zero.

It turns out that we can do a finer classification if we use *monoids*. Our endpoint is a classification where the differential finitely generated projective modules P and Q are equivalent when $P \oplus (R^n, D_0) \cong Q \oplus (R^m, D_0)$ for some m, n.

We need to use quotient monoids. Thus, let M be a monoid and N a submonoid. (By convention, submonoids have the same identity as the ambient monoid.) Suppose Q is a monoid and $p : M \to Q$ is a monoid homomorphism. (By convention, monoid homomorphisms carry identities to identities.) What does it mean if p sends every element of N to the identity? If m_1, m_2 are elements of M and n_1, n_2 are elements of N with $m_1 + n_1 = m_2 + n_2$ then since $p(n_i)$ is the identity, $p(m_1) = p(m_2)$.

As $\{(m_1,n_2)|p(m_1) = p(m_2)\}$ is an equivalence relation, we consider $\mathcal{E} = \{(m_1,n_2)|\exists n_1,n_2 \in N \text{ with } m_1+n_1 = m_2+n_2\}$ and see if it is a subequivalence relation. For any $m \in M$, $(m,m) \in \mathcal{E}$ with $n_1 = n_2 = 0$. If $(m_1,m_2) \in \mathcal{E}$ with n_1,n_2, then $m_2+n_2 = m_1+n_1$ so $(m_2,m_1) \in \mathcal{E}$ with n_2,n_1. Suppose $(m_1,m_2) \in \mathcal{E}$ with t_1,t_2 and $(m_2,m_3) \in \mathcal{E}$ with s_2,s_3 then $m_1+t_1 = m_2+t_2$ and $m_2+s_2 = m_3+s_3$. Adding s_2 to both sides of the first equation and t_2 to both sides of the second gives $m_1+t_1+s_2 = m_2+t_2+s_2$ and $m_2+s_2+t_2 = m_3+s_3+t_2$ so $m_1+(t_1+s_2) = m_2+t_2+s_2 = m_3+(s_3+t_2)$ so $(m_1,m_3) \in \mathcal{E}$ with $(t_1+s_2),(s_3+t_2)$. Thus $\mathcal{E}$ is symmetric and transitive, and satisfies the identity axiom, hence is an equivalence relation.

For convenience, we write $m_1 \sim m_2$ if $(m_1,m_2) \in \mathcal{E}$.Let M/N denote the set of equivalence classes of $\mathcal{E}$ on M. Suppose $m_1 m_2$ are in M with $m_1 \sim m_2$, say because $m_1+n_1 = m_2+n_2$. Let $m \in M$ and add it to both sides of the equation. Then $m_1+m+n_1 = m_2+m+n_2$ which shows that $m_1+m \sim m_2+m$. So if $m_{11},m_{12} \in M$ with $m_{11} \sim m_{12}$ and $m_{21},m_{22} \in M$ with $m_{21} \sim m_{22}$ then $m_{11}+m_{21} \sim m_{12}+m_{21}$ and $m_{21}+m_{12} \sim m_{22}+m_{12}$ which by transitivity means that $m_{11}+m_{21} \sim m_{21}+m_{22}$. This means that the equivalence classes M/N form a semigroup under addition such that $M \to M/N$ is additive. It is clear that the equivalence class of 0 is the identity for the semigroup M/N and hence that $M \to M/N$ is a monoid homomorphism. As we saw, if $p: M \to Q$ is a monoid homomorphism with $p(n) = 0\ \forall n \in N$ then p is constant on $\mathcal{E}$ equivalence classes so p factors through $M \to M/N$. We call M/N the *quotient* monoid of M by N. Obviously $M/M = \{0\}$. But consider the case where $M = \mathbb{Z}$ and $N = \{x \in \mathbb{Z}|x \geq 0\}$. For $a,b \in Z$ with $a-b \geq 0$ we have $a+0 = b+(a-b)$ so $a \sim b$ with $0, a-b \in N$. Thus $M/N = \{0\}$ even though $N \subsetneq M$.

Suppose M is a monoid and $K \subset N$ are submonoids. We can form the quotient monoid M/K and then restrict the homomorphism $M \to M/K$ to the submonoid N to get $N \to M/K$. This monoid homomorphism sends every element of K to the identity and hence induces a monoid homomorphism $N/K \to M/K$. We claim this map $p: N/K \to M/K$ is a monomorphism. Suppose the equivalence classes e_1, e_2 in N/K of the elements n_1 and n_2 of N go to the same element of M/K. By definition, $p(e_i)$ is the equivalence class to which n_i belongs. So $p(e_1) = p(e_2)$ means there are elements k_1, k_2 in K such that $n_1 + k_1 = n_2 + k_2$ so $n_1 \sim n_2$ and then $e_1 = e_2$.

5.2 MONOIDS OF PROJECTIVE MODULES

Throughout this section R is a commutative differential ring. We will be considering both finitely generated projective R modules and differential finitely generated projective R modules. To distinguish the latter case from the former will always apply the adjective "differential" or the superscript "diff". For example, as above, $M(R)$ denotes the set of isomorphism classes of finitely generated projective R modules, and $M^{\text{diff}}(R)$ denotes the set of isomorphism classes of differential finitely generated projective R modules. For a finitely generated projective (respectively differential finitely generated projective) R module P, let $[P]$ denote its isomorphism class. If we want to make the differential structure explicit, we will write (P, D_P) and $[(P, D_P)]$. We have commutative monoid structures on $M(R)$ and $M^{\text{diff}}(R)$ given by the operation $[P] + [Q] = [P \oplus Q]$ for which $[0]$ is an identity. Forgetting the differential structure defines a monoid homomorphism $M^{\text{diff}}(R) \to M(R)$ by $[(P, D_P)] \mapsto [P]$.

(To be a monoid homorphism a function must carry the identity to the identity and preserve the operation.) Because every finitely generated projective R module carries a differential structure, $M^{\text{diff}}(R) \to M(R)$ is surjective.

The group $K(M(R))$ generated by $M(R)$ is $K_0(R)$, and the group $K(M^{\text{diff}}(R))$ generated by $M^{\text{diff}}(R)$ is $K_0^{\text{diff}}(R)$. As before, for a finitely generated projective (respectively differential finitely generated projective) R module P, we write (P) for the image of $[P]$ in $K_0(R)$ (respectively $K_0^{\text{diff}}(R)$).

Let $F(R) \subseteq M(R)$ be the set of isomorphism classes of finitely generated free R modules; it is a submonoid of $M(R)$ since the direct sum of free modules is free (showing $F(R)$ is closed under the operation) and $[0] \in F(R)$ (showing the identity of $M(R)$ belongs to $F(R)$). Let $F^{\text{diff}}(R) \subseteq M^{\text{diff}}(R)$ be the set of isomorphism classes of differential finitely generated projective modules $[(P, D_P)]$ whose underlying module P is R free. The same reasoning shows that $F^{\text{diff}}(R)$ is a submonoid of $M^{\text{diff}}(R)$. We can also consider the subset $F_0^{\text{diff}}(R) \subset F^{\text{diff}}(R)$ consisting of isomorphism classes of the differential modules (R^n, D_0). This set is closed under direct sum $((R^n, D_0) \oplus (R^m, D_0) \cong (R^{n+m}, D_0))$ and contains the identity (R^0, D_0) and hence is a submonoid of $F^{\text{diff}}(R)$ and *a fortiori* of $M^{\text{diff}}(R)$.

The *projective class monoid* of R, denoted $PC(R)$, is the monoid quotient $M(R)/F(R)$. For a finitely generated projective module P, let (P) be the image of $[P]$ in the quotient monoid $PC(R)$. By definition, if P and Q are finitely generated projective R modules then $(P) = (Q)$ in $PC(R)$ if and only if there are n, m such that $P \oplus R^n \cong Q \oplus R^m$. For the case $P = R^n$, since $R^n \oplus \{0\} \cong \{0\} \oplus R^n$ says that $(R^n) = (\{0\}]$. Since for any finitely generated projective there is a finitely generated projective Q such that $P \oplus Q \cong R^n$ for some n in $PC(R)$ $(P)+(Q) = (\{0\})$ so $(P) \in U(PC(R))$. In other words, $PC(R)$ is a group to which $M(R)$ maps. Thus we have a (surjective) group homomorphism $K_0(R) \to PC(R)$. Note that (R) in $K_0(R)$ maps to $(R) = (\{0\})$ in $PC(R)$, which means $\mathbb{Z}(R)$ maps to $\{(\{0\})\}$ so the group homomorphism passes to a group homomorphism $\overline{K_0(R)} \to PC(R)$. On the other hand, the map $M(R) \to K_0(R)$ maps $F(R)$ into $\mathbb{Z}(R)$ so the composite $M(R) \to K_0(R) \to \overline{K_0(R)}$ is a monoid map which sends every element of $F(R)$ to the identity and hence factors through $PC(R) = M(R)/F(R)$. Consider the composites $PC(R) \to \overline{K_0(R)} \to PC(R)$ and $\overline{K_0(R)} \to PC(R) \to \overline{K_0(R)}$: for a finitely generated projective R module P the first composite is $(P) \mapsto (P) + \mathbb{Z}(R) \mapsto (P)$ and the second is $(P)+\mathbb{Z}(R) \mapsto (P) \mapsto (P)+\mathbb{Z}(R)$ showing that $PC(R) \to \overline{K_0(R)}$ and $\overline{K_0(R)} \to PC(R)$ are inverses. Thus $PC(R)$ is isomorphic to $\overline{K_0(R)}$. The lesson here is that while the K group $K_0(R)$ is not a quotient monoid of $M(R)$, the reduced K group $\overline{K_0(R)}$ is.

We can apply similar reasoning to the differential case. Let $PC^{\mathrm{diff}}(R)$ denote the monoid $M^{\mathrm{diff}}(R)/F^{\mathrm{diff}}(R)$. For a differential finitely generated projective R module (P, D_P), let $((P, D_P))$ denote the image of the isomorphism class $[(P, D_P)]$ in $PC^{\mathrm{diff}}(R)$. (When the context allows, we will refer to (P, D_P) as P as usual and den(te $((P, D_P))$ by (P)). For differential finitely generated projective modules P and Q, $(P) = (Q)$ then means that $P \oplus (R^n, D_A) \cong Q \oplus (R^m, D_B)$ for some n, m, sime $A \in M_n(R)$, and some $B \in M_m(R)$. In particular, since $\{0\} \oplus (R^n, D_A) \cong (R^n, D_A) \oplus (R^0, D_0)$, $((R^n, D_A)) = (\{0\})$ for any n and any $A \in M_n(R)$.

Proposition 5.2 *For any differential commutative ring R,*

$$PC^{diff}(R) = M^{diff}(R)/F^{diff}(R)$$

is a group. The induced group homomorphism

$$PC^{diff}(R) \to PC(R)$$

is an isomorphism.

Proof. Let (P, D_P) be any differential finitely generated projective R module. There is a finitely generated projective R module Q such that $P \oplus Q$ is free, say $P \oplus Q \cong R^n$. Put a differential structure D_Q on Q. Then $(P, D_P) \oplus (Q, D_Q)$ is a differential module, so $(P, D_P) \oplus (Q, D_Q) \cong (R^n, D_A)$ for some $A \in M_n(R)$. Thus $((P, D_P)) + ((Q, D_Q)) = ((R^n, D_A)) = (\{0\})$ in $PC^{\text{diff}}(R)$. Thus (Q) is an inverse to (P) in $PC^{\text{diff}}(R)$. Since every element of $PC^{\text{diff}}(R)$ has an inverse, $PC^{\text{diff}}(R)$ is a group. The monoid epimorphism $M^{\text{diff}}(R) \to M(R)$ gives a monoid epimorphism $M^{\text{diff}}(R) \to PC(R)$ which sends elements of $F^{\text{diff}}(R)$ to the identity and hence factors through $PC^{\text{diff}}(R)$. This monoid epimorphism $PC^{\text{diff}}(R) \to PC(R)$ is actually a group homomorphism. Suppose $((P, D_P))$ is in the kernel. Then $(P) = (\{0\})$ in $PC(R)$ which means that there are n, m such that $P \oplus R^n \cong \{0\} \oplus R^m$ as R modules. This means that there is an $A \in M_m(R)$ such that $(P, D_P) \oplus (R^n, D_0) \cong (R^m, D_A)$ so in $PC^{\text{diff}}(R)$ $((P, D_P)) = (\{0\})$. Thus the kernel is trivial and $PC^{\text{diff}}(R) \to PC(R)$ is an isomorphism.

We recall that $PC(R) \cong \overline{K_0(R)}$, thus Proposition 5.2 says $PC^{\text{diff}} \cong \overline{K_0)R)}$. We would like to relate this to the forget-the-derivation map $K_0^{\text{diff}}(R) \to K_0(R)$ induced from $((P, D_P)) \mapsto (P)$. We also know that

$\overline{K_0(R)}$ is isomorphic to the kernel of the rank map $\mathrm{rk} : K_0(R) \to \mathbb{Z}$. This map extends to differential finitely generated projective modules by defining $\mathrm{rk}(P, D_P)$ to be $\mathrm{rk}(P)$. (We are tacitly assuming that finitely generated projective R modules have constant rank.) Then $\mathrm{rk} : M^{\mathrm{diff}}(R) \to \mathbb{Z}$ is a monoid homorphism to a group and hence factors through $K_0^{\mathrm{diff}}(R)$; we use the same symbol so that $\mathrm{rk}(((P, D_P))) = \mathrm{rk}((P, D_P))$. Then we have a commutative diagram

$$\begin{array}{ccccccccc} 0 & \longrightarrow & \mathrm{Ker}^{\mathrm{diff}}(\mathrm{rk}) & \longrightarrow & K_0^{\mathrm{diff}}(R) & \xrightarrow{\mathrm{rk}} & \mathbb{Z} & \longrightarrow & 0 \\ \| & & \downarrow{\scriptstyle p} & & \downarrow{\scriptstyle q} & & \| & & \| \\ 0 & \longrightarrow & \mathrm{Ker}(\mathrm{rk}) & \longrightarrow & K_0(R) & \xrightarrow[\mathrm{rk}]{} & \mathbb{Z} & \longrightarrow & 0 \end{array}$$

(the superscript "diff" on the kernel in the first row is to indicate that it is the kernel of rk on $K_0^{\mathrm{diff}}(R)$). Since q is surjective, so is p. We know the bottom row is split by the map $\mathbb{Z} \to K_0(R)$ by $n \mapsto n(R)$. We can split the top sequence by $\mathbb{Z} \to K_0^{\mathrm{diff}}(R)$ by $n \mapsto n((R, D_0))$ and these are compatible:

$$\begin{array}{ccc} K_0^{\mathrm{diff}}(R) & \longleftarrow & \mathbb{Z} \\ \downarrow{\scriptstyle q} & & \| \\ K_0(R) & \longleftarrow & \mathbb{Z} \end{array}$$

commutes. To understand the kernel of q, we need only understand the kernel of p, which coincides with the former. So let $x \in K_0^{\mathrm{diff}}(R)$ be in the kernel of p. Using the canonical difference description we can write $x = ((P, D_P)) - ((Q, D_Q))$ for differential finitely generated projective modules (P, D_P) and (Q, D_Q), Since we assume $p(x) = 0$, we have that P and Q have the same rank m. There is a finitely generated projective R module M such that $Q \oplus M$ is free, say $Q \oplus M \cong R^n$. Then $(Q, D_M) \oplus (R^m, D_0) \cong (R^n, D_A)$ for some $A \in M_n(R)$. Since $((P, D_P)) + ((R^m, D_0) - ((Q, D_Q)) + ((R^m.D_0))) = ((P, D_P) - ((Q, D_Q))$, we can assume initially that $x = ((P, D_P) - ((R^n, D_A))$ where P has rank n. To say that x is zero in $K_0^{\text{diff}}(R) = K(M^{\text{diff}}(R))$ means that there is a differential finitely generated projective module T such that $(P, D_P) \oplus (T, D_T) \cong (R^n, D_A) \oplus (T, D_T)$. By adding a suitable finitely generated projective R module on which we have placed a differential structure, we may assume T is R free so that $(P, D_P) \oplus (R^p, D_B) \cong (R^n, D_A) \oplus (R^p, D_B)$. The point is that the same element of $F^{\text{diff}}(R)$ has been added to both sides. To say that x is in the kernel of q means that $(P) - (R^n)$ is zero in $K_0(R)$ so $(P) = (R^n)$ in $K_0(R)$. Since, by definition, $K_0(R) = K(M(R))$ this means there is a finitely generated projective R module M such that $P \oplus M \cong R^n \oplus M$, and then adding an appropriate module N to both sides such that $M \oplus N$ is free, say $M \oplus N \cong R^m$, we ultimately have $P \oplus R^m \cong R^n \oplus R^m$. This is the formulation that $p(x)$ is zero. If we can find a matrix C such that $(P, D_P) \oplus (R^m, D_C) \cong (R^n, D_A) \oplus (R^m, D_C)$ then we can conclude that x is zero.

Such a C may not exist, however. Consider the case $R = \mathbb{C}$ and $D_R = 0$. Since all $\mathbb{C}$ modules are vector spaces, differential finitely generated projective modules (P, D_P) are all of the form $(\mathbb{C}^M, D_A)$. If $S \in GL_n(\mathbb{C})$ is a differential map from $(\mathbb{C}^n, D_A) \to (\mathbb{C}^n, B)$ then (since $S' = 0$) $B = SAS^{-1}$. When $n = 1$, then this reduces to the statement that $(R^1, D_A) \cong (R^1, D_B)$ if and only if $A = B$. Suppose $A = [a]$ and $B = [b]$, and let C be any matrix. Let $\text{ch}(Y)(x)$ be the characteristic polynomial of the matrix Y in the variable x. Then for any matrix $m \times m$ C, $\text{ch}(A \oplus C) = (x - a)\text{ch}(C)$ and $\text{ch}(B \oplus C) = (x - b)\text{ch}(C)$ so $(\mathbb{C}^1, D_A) \oplus (\mathbb{C}^m, D_C) \cong (\mathbb{C}^1, D_B) \oplus (\mathbb{C}^m, D_C)$ if and only if $a = b$. Choose $a \neq b$. Then $x((\mathbb{C}^1, D_A)) - ((\mathbb{C}^1, D_B)) \neq (\{0\})$ in $\text{Ker}^{\text{diff}}(\text{rk})$ (hence also non-zero in $K_0^{\text{diff}}(\mathbb{C})$ but $p(x) = (\mathbb{C}) - (\mathbb{C}) = (\{0\})$. This same example shows that the surjection $K_0^{\text{diff}}(\mathbb{C}) \to K_0(\mathbb{C})$ is not an isomorphism.

To summarize this discussion: Proposition 5.2 shows that differential projective modules up to ("modulo") differential modules which are R free depend only on the underlying projective module up to free modules. Thus the classification of differential modules might as well start with those which are free as R modules. We have called the monoid of isomorphism classes of differential finitely generated R modules whose underlying module is free $F^{\text{diff}}(R)$ and the submonoid of $F^{\text{diff}}(R)$ of isomorphism classes of the modules (R^n, D_0) $F_0^{\text{diff}}(R)$. We intend to focus on the monoid $F^{\text{diff}}(R)/F_0^{\text{diff}}(R)$ (denoted henceforth $FC_0^{\text{diff}}(R)$) which describes the differential modules free as R modules modulo those whose derivation is diagonal. As noted, Proposition 5.2 can be understood as implying that $FC_0^{\text{diff}}(R)$ is the heart of classifying the differential finitely generated projective R modules, at least in the spirit of K theory. Further in that spirit, since what we are regarding as trivial are the free modules with diagonal derivation, the differential analogue of K_0 should be $PC_0^{\text{diff}}(R) = M^{\text{diff}}(R)/F_0^{\text{diff}}(R)$.[2]

In case every differential module of the form (R^n, D_A) has "stably diagonal derivation" in the sense that $(R^n, D_A) \oplus (R^m, D_0) \cong (R^{n+m}, D_0)$ then $FC_0^{\text{diff}}(R)$ is trivial so $PC_0^{\text{diff}}(R) = PC^{\text{diff}}(R)$ which is isomorphic to $\overline{K_0(R)}$, reinforcing the idea that $FC_0^{\text{diff}}(R)$ captures the differential heart of the classification of differential modules.

If $\overline{K_0(R)}$ is trivial, so that all finitely generated projective R modules are stably free, then $FC_0^{\text{diff}}(R)$ should classify all differential finitely generated projective modules. This is indeed the case:

---

[2]In Differential Projective Modules over Differential Rings, II $PC_0^{\text{diff}}(R)$ was denoted $PC(R)$ and $PC(R)$ was denoted $PC(R)/FC^{\text{diff}}(R)$.

Proposition 5.3 *Suppose all finitely generated projective R modules are stably free. Then* $FC_0^{diff}(R) = PC_0^{diff}(R)$.

Proof. The hypothesis asserts that for any finitely generated projective R module P, there are m, n such that $P \oplus R^n \cong R^m \oplus R^n$. Suppose P has a differential structure D_P. Consider the differential module $(P, D_P) \oplus (R^n, D_0)$. Since $P \oplus R^m$ is isomorphic to the free module R^{m+n}, there is a matrix A such that $(R^{m_n}, D_A) \cong (P, D_P) \oplus (R^n, D_0)$. Thus in $PC_0^{\text{diff}}(R) = M^{\text{diff}}(R)/F_0^{\text{diff}}(R)$,$((P, D_P)) = ((R^n, D_A))$. Since $((R^n, D_A))$ belongs to the submonoid $FC_0^{\text{diff}}(R) = F^{\text{diff}}(R)/F_0^{\text{diff}}(R)$, $((P, D_P)) \in FC_0^{\text{diff}}(R)$.

All of the monoids which we have introduced in this section are functorial in R: ring homomorphisms $R \to S$, or differential ring homomorphisms $R \to S$ when S is a differential ring, induce monoid homomorphosms.

5.3 UNITS

We can pass from differential finitely generated projective R modules to finitely generated projective R modules by forgetting the differential structure. In general there is no obvious converse (a way to put a canonical differential structure on finitely generated projective modules); even in the case of free modules, we still have to choose a basis to choose a matrix to give a differential structure. There is, however, a natural differential R module structure on modules induced from R^D via action on R: namely, acting on $R \otimes_{R^D} M$ via the differential action on the first factor R. And if M is a finitely generated projective R^D module, $R \otimes_{R^D} M$ is a differential finitely generated projective R module. Moreover, the classes in $PC_0^{\text{diif}}(R)$ of such modules can be identified:

Proposition 5.4 *The functor $P \mapsto R \otimes_{R^D} P$ from finitely generated projective R^D modules to differential finitely generated projective R modules induces an isomorphism*

$$\overline{K_0(R^D)} \to U(PC_0^{\textit{diff}}(R)).$$

Proof. For proof purposes, it is convenient to deal with $\overline{K_0(R^D)}$ via its isomorphism with $PC(R^D)$. Because the tensor product distributes over the direct sum (up to isomorphism), and induction preserves finite generation and projectivity, the functor induces a monoid homomorphism $M(R^D) \to M^{\text{diff}}(R)$. If P is finitely generated and R^D free, so is $R \otimes_{R^D} P$. As a differential module, $R \otimes_{R^D} (R^D)^n$ is isomorphic to (R^n, D_0), Thus the monoid homomorphism carries $F(R^D)$ to $F_0^{\text{diff}}(R)$ and therefore gives a monoid map $(P) \mapsto (R \otimes_{R^D} P)$. Suppose P is a finitely generated projective R^D modules, and let Q be a finitely generated projective R^D module such that $P \oplus Q$ is free, say of rank n. Then in $PC_0^{\text{diff}}(R)$, $(R \otimes_{R^D} P) + (R \otimes_{R^D} Q) = (R \otimes_{R^D} (P \oplus Q)) = ((R^n, D_0)) = (\{0\})$. Thus the monoid homorphism is actually a group homomorphism $\psi : PC(R^D) \to U(PC_0^{\text{diff}}(R))$.

To see that ψ is an isomorphism, we first check surjectivity. So suppose M is a differential finitely generated projective R module such that (M) is a unit of $PC_0^{\text{diff}}(R)$, say with inverse (N), where N is also a differential finitely generated R module. To say $(M) + (N) = (\{0\})$ means that there are m, n such that $M \oplus N \oplus (R^n, D_0) \cong \{0\} \oplus (R^m, D_0)$ differentially. This means that M is a differential direct summand of (R^n, D_0).

By Theorem 4.3 of Chapter this means that as a differential R module, M is isomorphic to $R \otimes_{R^D} P$ for some finitely generated projective R^D module P. The existence of P for M shows that ψ is surjective.

To see that ψ is injective, suppose (P) is in its kernel. That is $(R \otimes_{R^D} P) = (\{0\})$. Then there are m, n such that there is a differential isomorphism $(R \otimes_{R^D} P) \oplus (R^n, D_0) \cong \{0\} \oplus (R^m, D_0)$. We have differential isomorphisms $(R^n, D_0) \cong R \otimes_{R^D} (R^D)^n$ and $(R^m, D_0) \cong R \otimes_{R^D} (R^D)^m$, so also we have differential isomorphisms $(R \otimes_{R^D} P) \oplus (R^n, D_0) \cong R \otimes_{R^D} (P \oplus (R^D)^n)$. Taking constants of the last two gives $(R \otimes_{R^D} (R^D)^m)^D = (R^D)^m$ and $R \otimes_{R^D} ((P \oplus (R^D)^n)^D = P \oplus (R^D)^n$. Since $(R \otimes_{R^D} (R^D)^m) \cong R \otimes_{R^D} ((P \oplus (R^D)^n$ this means that $(R^D)^m \cong P \oplus (R^D)^n$. Thus in $PC(R^D)$, $(P) = (\{0\})$. Thus the kernel of ψ is trivial, so ψ is an isomorphism.

The functor $P \mapsto R \otimes_{R^D} P$ is also a functor from finitely generated projective R^D modules to finitely generated projective R modules which, due to its additivity, induces a homomorphism of reduced K groups $\overline{K_0(R^D)} \to \overline{K_0(R)}$. By Proposition 5.4 we know the domain is isomorphic to $U(PC_0^{\text{diff}}(R))$. Under that isomorphism we can identify its kernel.

Lemma 5.1 *The kernel of* $U(PC_0^{diff}(R)) \cong \overline{K_0(R^D)} \to \overline{K_0(R)}$ *is* $U(FC_0^{diff}(R))$

Proof. The kernel of $\overline{K_0(R^D)} \to \overline{K_0(R)}$ consists of the classes of finitely generated projective R^D modules P such that $R \otimes_{R^D} P$ is a stably free R module. That is, there are n, m such that $(R \otimes_{R^D} P) \oplus R^n \cong R^m$. Since $(R \otimes_{R^D} P) \oplus R^n \cong R \otimes_{R^D} (P \oplus (R^D)^n)$ and $(P) = (P \oplus (R^D)^n)$ in $\overline{K_0(R^D)}$, we can replace P with $P \oplus (R^D)^n$ and assume that $R \otimes_{R^D} P \cong$

R^m. Thus $R \otimes_{R^D} P$ is R free. If we think of it as a differential module, then it lies in the subset $FC_0^{\text{diff}}(R) \subseteq PC_0^{\text{diff}}(R)$. By Proposition 5.4, it also lies in $U(PC_0^{\text{diff}}(R))$ and hence in $FC_0^{\text{diff}}(R) \cap U(PC_0^{\text{diff}}(R))$. Thus if P is a finitely generated projective R^D module with $R \otimes_{R^D} P$ stably free, then the class of this latter lies in this intersection. On the other hand, suppose we have a class in $PC_0^{\text{diff}}(R)$ which lies in the intersection. Then on the one hand, it is of the form (R^n, D_A) and on the other of the form $(R \otimes_{R^D} P)$. This means there is an m such that $IR^n, D_A) \oplus (R^m, D_0) \cong (R \otimes_{R^D} P) \oplus (R^m, 0)$. Thus $(R^{n+m}, D_B) \cong (R \otimes_{R^D} P) \oplus (R^m, D_0)$ where $B = A \oplus 0$, so on the underlying projective modules $R^{n+m} \cong (R \otimes_{R^D} P) \oplus R^m$ so $R \otimes_{R^D} P$ is stably free. Thus the intersection is the kernel of $U(PC_0^{\text{diff}}(R)) \cong \overline{K_0(R^D)} \to \overline{K_0(R)}$. It remains to show that $U(FC_0^{\text{diff}}(R) = FC_0^{\text{diff}}(R) \cap U(PC_0^{\text{diff}}(R))$. We clearly have the inclusion $U(FC_0^{\text{diff}}(R) \subseteq FC_0^{\text{diff}}(R) \cap U(PC_0^{\text{diff}}(R))$. On the other hand, we know the intersection is the kernel of a group homorphism, and hence itself a group. Since it is a submonoid of $FC_0^{\text{diff}}(R)$, it is contained in $U(FC_0^{\text{diff}}(R))$. We conclude equality and hence the lemma.

The homomorphism in Lemma 5.1 is a morphism of K groups, which has nothing to do with R being a differential ring. Thus perhaps it is surprising that if $U(FC_0^{\text{diff}}(R))$ is trivial then $\overline{K_0(R^D)} \to \overline{K_0(R)}$ is injective. In fact, we have

Corollary 5.2 *Suppose $FC_0^{diff}(R)$ is trivial. Then $\overline{K_0(R^D)} \to \overline{K_0(R)}$ is an isomorphism.*

Proof. When $FC_0^{\text{diff}}(R)$ is trivial $PC_0^{\text{diff}}(R) = PC^{\text{diff}}(R)$. By Proposition 5.2, $PC^{\text{diff}}(R) = \overline{K_0(R)}$ is a group, which means that

$U(PC^{\text{diff}}(R)) = PC^{\text{diff}}(R)$. By Proposition 5.4 $\overline{K_0(R^D)} = U(PC_0^{\text{diff}}(R))$. Thus $\overline{K_0(R^D)} = U(PC_0^{\text{diff}}(R)) = U(PC^{\text{diff}}(R)) = PC^{\text{diff}}(R) = \overline{K_0(R)}$.

The condition of Corollary 5.2 – that $FC_0^{\text{diff}}(R) = 0$ – means that every differential R module of the form (R^n, A) is equivalent to one of the form $(R^m, 0)$, where "equivalent" here means for erey n, and for every $A \in M_n(R)$ there is m such that $(R^n, D_A) \oplus (R^m, D_0) \cong (R^{n+m}, D_0)$. If the matrix $S \in GL_{n+m}(R)$ gives the isomorphism then $A \oplus 0 = S^{-1}S'$. We consider what this means in the case $n = 1$ and $A = [a]$. Then there is an m such that $B = A \oplus \mathbf{0}_m$ (where $\mathbf{0}_m$ is the $m \times m$ zero matrix) satsifies $S' = SB$ for some invertible $(1+m) \times (1+m)$ matrix $S = [s_{ij}]$. The matrix SB has first column $[as_{11}as_{21}\ldots]^T$ and all other entries zero so $S' = SB$ means that $s'_{i1} = as_{i1}$ for $1 \leq i \leq 1+m$ and $s'_{ij} = 0$ for $j \neq 1$. Thus if σ is any permutation of $\{1, 2, \ldots, 1+m\}$,

$$(s_{\sigma(1)1}s_{\sigma(2)2}\cdots)' = (as_{\sigma(1)1}s_{2\sigma(2)}\cdots).$$

Thus $u = \det(S)$ satisfies $u' = au$. Since u is a unit, we conclude that if there is an m such that $(R^1, D_A) \oplus (R^m, D_0) \cong (R^{1+m}, D_0)$ then there is a unit u such that multiplication by u is a differential isomorphism $(R^1, D_{[a]}) \to (R^1, D_0)$. So the condition that $FC_0^{\text{diff}}(R)$ is trivial implies that every element of R is the logarithmic derivative of a unit.

One case where the latter does not obtain is $R = \mathbb{C}[x]$ with $D(x) = 1$ and $D(c) = 0$ for $c \in \mathbb{C}$. In this case the units are all constants and the only logarithmic derivative is 0, so for example $((R^1, D_{[x]}))$ is a non trivial element of $FC_0^{\text{diff}}(R)$. On the other hand, since all finitely generated projective $R = \mathbb{C}[x]$ modules are free, as are all $R^D = \mathbb{C}$ modules, their reduced K groups are trivial and hence isomorphic. That is, for

$R = \mathbb{C}[x]$, $\overline{K_0(R^D)} \to \overline{K_0(R)}$ is an isomorphism, but $FC_0^{\text{diff}}(R)$ is not trivial, showing that the converse of Corollary 5.2 doesn't hold.

We examine a consequence of Proposition 5.4:

Corollary 5.3 *Suppose that $\overline{K_0(R^D)}$ is trivial. Then the only unit of $PC_0^{diff}(R)$ is the identity.*

If $PC_0^{\text{diff}}(R)$ is trivial, then so is $PC^{\text{diff}}(R)$; since the latter is isomorphic to $\overline{K_0(R)}$ by Proposition 5.2 then the reduced K_0 of R is also trivial. And if $PC_0^{\text{diff}}(R)$ is trivial, then so is its submonoid $FC_0^{\text{diff}}(R)$. By Corollary 5.2 this means that $\overline{K_0(R^D)}$ is isomorphic to $\overline{K_0(R)}$ so $\overline{K_0(R^D)}$ is trivial also.

To say that $PC_0^{\text{diff}}(R)$ is trivial means that if (P, D_P) is a differential finitely generated projective R module then $((P, D_P)) = ((R^n, D_0))$ for some n, which means that (D, D_P) and (R^n, D_0) are isomorphic, after the addition of (R^m, D_0) to both, Similarly, $\overline{K_0(R)}$ is trivial, and P is a finitely generated projective R module then $(P) = (R^n)$ in $K_0(R)$ (for some n); this means that P and R^n are isomorphic – after the addition to both of some free module.

A stronger (unstabilized) version of these facts is also true.

Lemma 5.2 *Suppose any differential finitely generated projective R module is isomorphic to one of the form (R^n, D_0). Then all finitely generated projective R modules and all finitely generated projective R^D modules are free.*

Proof. Assume the hypothesis. If P is a finitely generated projective R module, then there exists a differential structure D_P on P. Then there is a differential isomorphism $(P, D_P) \cong (R^n, D_0)$ for some n, and therefore an R module isomorphism $P \cong R^n$. If Q is a finitely generated projective R^D module, then $R \otimes_{R^D} Q$ is a differential finitely generated projective R module. There is then a differential isomorphism $R \otimes_{R^D} Q \cong (R^n, D_0)$ (for some n) and then $P \cong (R \otimes_{R^D} P)^D \cong (R^n, D_0)^D = (R^D)^n$.

5.4 TRIVIAL $PC_0^{\mathrm{DIFF}}(R)$

The condition that $PC_0^{\mathrm{diff}}(R)$ is trivial is the condition that after the addition of a differential module of the form (R^n, D_0) every differential module is isomorphic to one of this latter form, This is a stable form of the stronger condition that every differential finitely generated projective R module is isomorphic to (R^n, D_0) for some n, and is the best we can hope for working with differential projective class monoids where the modules of the form (R^n, D_0) are declared trivial ($PC_0^{\mathrm{diff}}(R) = M^{\mathrm{diff}}(R)/F_0^{\mathrm{diff}}(R)$). The unstable version – that every differential finitely generated projective R module has the form (R^n, D_0) certainly happens; this is the assertion that every differential finitely generated projective R module has a basis of constants. Typically, however, we look at the stable version and then see if the unstable can be deduced from it by means of some sort of cancellation theorem. We will look at cancellation in a later section. In this one, we concentrate on the consequences of trivial $PC_0^{\mathrm{diff}}(R)$.

One such consequence is that the submonoid $FC_0^{\mathrm{diff}}(R)$ is trivial. Consider this condition by itself. That is, for any n, and $A \in M_n(R)$ there are p, q such that $(R^n, D_A) \oplus (R^p, D_0) \cong (R^q, D_0)$. (Technically, the right-hand side here is $(R^0, D_0) \oplus (R^q, D_0)$.) Now suppose (P, D_P) is any differential finitely generated projective R module. Let Q be a finitely generated projective R module such that $P \oplus Q$ is a free R module, say $P \oplus Q \cong R^n$. Let D_Q be a differential structure on Q. Then $(P, D_P) \oplus (Q, D_Q) \cong (R^n, D_A)$ for some $A \in M_n(R)$. Let p, q be as above. Then we have $(P, D_P) \oplus (Q \oplus R^p, D_{A \oplus 0}) \cong (R^q, D_0)$. This shows that (P, D_P) is invertible in $PC_0^{\text{diff}}(R)$, and therefore in the image of $\overline{K_0(R^D)}$. And if this latter is trivial then so is $PC_0^{\text{diff}}(R)$.

Proposition 5.5 *If $FC_0^{diff}(R)$ is trivial then $\overline{K_0(R^D)} \to PC_0^{diff}(R)$ is surjective. If also $\overline{K_0(R^D)}$ is trivial then so is $PC_0^{diff}(R)$.*

The condition that $\overline{K_0(R)}$ is trivial means that all finitely generated projective R^D modules are stably free. Suppose the stronger condition: that all finitely generated projective R^D modules are free:

Lemma 5.3 *Suppose all finitely generated projective R^D modules are free, Let (P, D_P) be a differential finitely generated projective R module where P has rank n such that there are p, q such that $(P, D_P) \oplus (R^p, D_0) \cong (R^q, D_0)$. Then $(P, D_P) \cong (R^{q-p}, D_0)$.*

Proof. By hypothesis (P, D_P) is a differential direct summand of (R^q, D_0). By Proposition 4.3 $(P, D_P) \cong R \otimes_{R^D} Q$ where Q is a finitely generated projective R^D module, of rank n since P has rank n. By assumption, Q is free. Therefore, since $R \otimes_{R^D} (R^D)^n \cong (R^n, D_0)$ and $n + p = q$, the result follows.

So far, we have not used the funtorality of PC_0^{diff} or FC_0^{diff}. But this can come in handy. We will be particularly interested in the case where we have a *surjective* differential ring homomorphism $f : R \to S$. In this case $M_n(R) \to M_n(S)$ is surjective for every n, so given $A \in M_n(S)$, we can find $B \in M_n(R)$ with $M_n(f)(B) = A$. This means that $(S^n, D_A) \cong S \otimes_f (R^n, D_B)$ which implies that the induced monoid homomorphism $FC^{\text{diff}}(R) \to FC_0^{\text{diff}}(S)$ is surjective. While we do not know anything in general about the surjectivity of $PC_0^{\text{diff}}(R) \to PC_0^{\text{diff}}(S)$, we do have the following:

Lemma 5.4 *Let $f : R \to S$ be a surjective differential ring homomorphism. Assume $\overline{K_0(S^D)}$ is trivial. If $PC_0^{diff}(R)$ is trivial, $PC_0^{diff}(S)$ is trivial.*

Proof. The submonoid $FC_0^{\text{diff}}(R)$ of the trivial monoid $PC_0^{\text{diff}}(R)$ is also trivial, so the fact that $FC_0^{\text{diff}}(R) \to FC_0^{\text{diff}}(S)$ is surjective implies that $FC_0^{\text{diff}}(S)$ is trivial. Since we also hypothesize that $\overline{K_0(S^D)}$ is trivial, by Proposition 5.5 we have that $PC_0^{\text{diff}}(S)$ is trivial.

Suppose R is a differential ring with trivial $PC_0^{\text{diff}}(R)$ and I is a differential ideal of R. Lemma 5.4 asserts that if $\overline{(K_0((R/I)^D)}$ is trivial so is $PC_0^{\text{diff}}(R/I)$. Now if I is a maximal differential ideal, R/I is a simple differential ring. Any element of $(R/I)^D$ generates a differential principal ideal of R/I; since R/I is simple the differential ideals are R/I and $\{0\}$ so any non–zero element of $(R/I)^D$ is a unit and $(R/I)^D$ is a field. In particular, $\overline{K_0((R/I)^D)}$ is trivial. Thus we have the following corollary to Lemma 5.4:

Corollary 5.4 *Let R be a differential ring with $PC_0^{diff}(R)$ trivial. Then for every maximal differential ideal I of R, $PC_0^{diff}(R/I)$ is trivial.*

The main use of Corollary 5.4 will be its contrapositive to exhibit examples of rings with non–trivial PC_0^{diff}.

5.5 PICARD–VESSIOT EXAMPLES

Suppose F is a differential field of characteristic zero with an algebraically closed field of constants C. All F modules are projective, in fact free, of course. As in Chapter 2, we fix a Picard–Vessiot closure F^{PV} of F. For every finite-dimensional differential F module V, there is a unique subfield $E_V \subseteq F^{PV}$ which is minimal with respect to the condition that $E_V \otimes_F V$ is isomorphic to (E_V^n, D_0) as differential E_V modules. In addition, E_V is a Picard–Vessiot extension of F (and every Picard–Vessiot extension is of this form).

Suppose V and W are finitely generated differential F modules of dimension n and m with $(V) = (W)$ in $PC_0^{\mathrm{diff}}(F)$. That is, there are p, q such that $V \oplus (F^p, D_0) \cong W \oplus (F^q, D_0)$. Since $E_V \otimes V \cong (E_V^n, D_0)$, $E_V \otimes (V \oplus (F^p, D_0)) \cong (E_V^{n+p}, D_0)$ which implies that $E_V \otimes (W \oplus (F^q, D_0)) \cong (E_V \otimes W) \oplus (E_V^q, D_0) \cong (E_V^{m+q}, D_0)$. Since $(E_V)^D = C$ is a field, Then Lemma applied to $R = E_V$ and $P = E_V \otimes W$ implies that $E_V \otimes W \cong (E_V^m, D_0)$. Thus E_V is a splitting field, contained in F^{PV}, for W, so $E_W \subseteq E_V$. Similar reasoning applied to W shows that $E_V \subset E_W$. We conclude that $E_V = E_W$. Thus $(V) = (W)$ implies that $E_V = E_W$. This has the following consequence:

Proposition 5.6 *Let F be a differential field such that $PC_0^{diff}(F)$ is finite. Then F^{PV} is a (finitely generated) Picard–Vessiot extension of F. If $PC_0^{diff}(F)$ is trivial then $F^{PV} = F$.*

Proof. Suppose $PC_0^{\mathrm{diff}}(F) = \{(V_1), (V_2) \ldots, (V_n)\}$. Let $E_i = E_{V_i}$, and let E be the compositum of $E_1, \ldots E_n$. Note: E is a Picard–Vessiot extension of F. Let V be a finite-dimensional differential F module.

Since $(V) = (V_i)$ for some i, $E_V = E_i$ for that i. Thus $E_V \subseteq E$. Hence F^{PV}, which is the compositium of all E_V for all V, is acontained in, hence equals, its subfield E. Note that F is the splitting field for (F^n, D_0) for any n, including $n = 0$. So if $PC_0^{\text{diff}}(F) = \{(\{0\})\}$, $E = F$ and $F^{PV} = F$.

Usually F^{PV} is not a (finitely generated) Picard–Vessiot extension of F. One way to express this is to say that F has Picard–Vessiot extensions of arbitrarly large transcendence degree. In tis case. $PC^{\text{diff}}(F)$ must be infinite.

Corollary 5.5 *Suppose for the differential F, of characteristic zero and with an algebraically closed field of constants, the transcendence degree of its Picard–Vessiot extensions is unbounded. Then $PC_0^{diff}(F)$ is infinite. Furthermore, if $K \supseteq F$ is a Picard–Vessiot extension, $PC_0^{diff}(K)$ is infinite.*

Proof. If E is a Picard–Vessiot extension of F, the compositum $EK \subset F^{PV}$ is a Picard–Vessiot extension of K. If all these have bounded transcendence degree over K, they have bounded transcendence degree over F, contrary to the hypothesis.

Now suppose F is a differential field of characteristic zero with an algebraically closed field of constants C which has Picard–Vessiot extensions of arbitrarily large transcendence degree, and that $R \supseteq F$ is a differential R algebra which is a finitely generated F algebra. Let I be a maximal differential ideal of R. Then R/I is an integral domain, and its quotient field K is a finitely generated differential field extension of F with $K^D = C$. Since R is finitely generated as an R algebra, the transcendence degree of $K \supseteq F$ is finite. This means that there are Picard–Vessiot extensions of F whose transcendence degree is so large they can't be differentially embedded in K over F. Let V be one such finite dimensional differential F module, say of dimension n, such that E_V has no differential image in K. If the differential E module $E \otimes_F V$ were isomorphic to (E^n, D_0) as differential modules over E, $E \otimes_F V$ would have a basis of constants over E, and the differential subfield of E generated by the elements of E necessary to describe those basis elements as tensors would be a minimal splitting field inside K for V over F, which, since it is a Picard–Vessiot extension, would have to be differentially isomorphic to E_V over F. We conclude that $E \otimes_F V$ can't be isomorphic to (E^n, D_0). This means that $R/I \otimes_F V$ can't be isomorphic to $((R/I)^n, D_0)$ either. Since $(R/I)^D$ is the field C, all its finitely generated projective modules are free. Thus we can apply Lemma 5.2 to conclude that $(V) \neq (\{0\})$ in $PC_0^{\text{diff}}(R/I)$. Then Corollary 5.4 implies that $PC_0^{\text{Diff}}(R)$ is not trivial.

To summarize:

Proposition 5.7 *Let F be a differential field of characteristic zero with an algebraically closed field of constants. Assume that the transcendence degrees of Picard–Vessiot extensions are not bounded. Let R be a finitely generated differential F algebra. Then $PC_0^{diff}(R)$ is not trivial.*

There are many fields F which satisfy the hypotheses of Proposition 5.7, for example, the constant field $\mathbb{C}$ and the rational functional fields $\mathbb{C}(t_1, \dots, t_n)$. Then any finitely generated differential F algebra R has to have at least one differential finitely generated projective module (P, D_P) such that for no p, n is $(P, D_P) \oplus (R^q, D_0) \cong (R^n, D_0)$. Here R can be a polynomial ring $F[x_1, \dots x_n]$ with derivation given by any n-tuple of polynomials $(f_1, \dots, f_n)$ where $D_R(x_i) = f_i$ (and $D_R(a) = D_F(a)$ for $a \in$

F). This suggests that it is rare for differential rings to have a trivial PC_0^{diff}, although we have seen examples (R^{PV} and $\mathcal{O}$) where this occurs.

5.6 CANCELLATION

A monoid M has *cancellation* provided $m+p=n+p$ implies $m=n$ for any elements m,n,p of M. This happens, of course: it's always true in a group, and its true in monoids like the non–negative integers, and direct products of monoids with cancellation have cancellation. In terms of the concept of cancellation, according to Proposition 5.1, for any monoid M, $M \to K(M)$ is injective if and only if M has cancellation. Or maybe a given monoid M does not have cancellation, but it might have it with respect to a submonoid N: given p,q,n in M with $n \in N$ then $p+n=q+n$ implies $p=q$. For a commutative ring R, a projective module P is stably free if there are m,n such that $P \oplus R^m \cong R^n$. If P is free, say $P = R^k$, then P is stably free (for instance, $R^k \oplus R^1 \cong R^{k+1}$, but the converse fails: there are examples of stably free modules which are not free. Suppose P as above is stably free of rank k, so $n=k+m$ so $P \oplus R^m \cong R^k \oplus R^m$. Thus $[P]+[R^m]=[R^k]+[R^m]$ in the monoid $M(R)$ of isomorphism classes of finitely generated projective R modules. If $M(R)$ has cancellation then $[P]=[R^k]$ so $P \cong R^k$. In any event the group $K_0(R)$ has cancellation so $(P)=(R^k)$ in $K_0(R)$. If P is not free, then P being stably free is an example of the failure of cancellation. In this example the element of $M(R)$ that can't be cancelled is $[R^m]=m[R]$, which is an element of the submonoid $N=\{m[R] | m \geq 0\}$. It is a trivial induction argument to see that cancellation with respect to N is equivalent to just being able to cancel $[R]$. It is possible that R has no non-free stably free modules but

$M(R)$ still doesn't have cancellation with respect to $[R]$: there might be non-isomorphic P,Q with $P \oplus R^m \cong Q \oplus R^m$. (The best we can conclude from all stably free are free is that for any finitely generated projective R module L if $P \oplus L$ is free then so is $Q \oplus L$.) Nonetheless, cancellation of $[R]$ does imply that stably free are free.

We want to look at similar concepts for differential finitely generated projective R modules. Here the analogue of cancelling $[R]$ will be cancelling $[(R^1, D_0)]$. We say that the differential ring R has *free cancellation* if for differential finitely generated projective R modules (P, D_P) and (Q, D_Q) $(P, D_P) \oplus (R^1, D_0) \cong (Q, D_Q) \oplus (R^1, D_0)$ implies $(P, D_P) \cong (Q, D_Q)$. Again, a trivial induction says that if R has free cancellation then $(P, D_P) \oplus (R^k, D_0) \cong (Q, D_Q) \oplus (R^k, D_0)$ implies $(P, D_P) \cong (Q, D_Q)$. And this latter says that if $m \geq n$, $(P, D_P) \oplus (R^n, D_0) \cong (Q, D_Q) \oplus (R^m, D_0)$ implies $(P, D_P) \cong (Q, D_Q) \oplus (R^{m-n}, D_0)$. Since $((P, D_P)) = ((Q, D_Q))$ in $PC_0^{\text{diff}}(R)$ is precisely the statement that $(P, D_P) \oplus (R^n, D_0) \cong (Q, D_Q) \oplus (R^m, D_0)$ for some m, n. Either $m \geq n$ or $n \geq m$, so in the case for free cancellation either $(P, D_P) \cong (Q, D_Q) \oplus (R^{m-n}, D_0)$ or $(Q, D_Q) \cong (Q, D_Q) \oplus (R^{n-m}, D_0)$. Notice that $m - n$ or $n - m$ can be computed from the ranks of P and Q.So if R has free cancellation and $((P, D_P)) = ((Q, D_Q))$ in $PC_0^{\text{diff}}(R)$ with P of rank p and Q of rank q then $(P, D_P) \cong (Q, D_Q) \oplus (R^{p-q}, D_0)$ for $p \geq q$ or $(Q, D_Q) \oplus (R^{q-p}, D_0)$ if $q \geq p$. If $p = q$ then $((P, D_P) \cong (Q, D_Q)$.

In other words, given free cancellation, equality in $PC_0^{\text{diff}}(R)$ is the same as isomorphism, with copies of (R^1, D_0) added as necessary to equalize rank.

Free cancellation is a strong condition, but it happens in many of the examples we have considered, as the following proposition shows:

Proposition 5.8 *If R^D is a field, R has free cancellation.*

Proof. Let C denote the field R^D. We want to show that if P and Q are differential finitely generated projective modules with $P \oplus (R^1, D_0) \cong Q \oplus (R^1, D_0)$ then $P \cong Q$. We set up the situation as follows. Let M be a differential finitely generated projective R module. Write $M \oplus (R^1, D_0)$ as $M \times R = \{(m,r) | m \in M \text{ and } r \in R\}$ with derivation $D((m,r)) = (D_M(m), D_R(r))$. Then suppose we have differential submodules $N_1 \subset M \times R$ and $R_1 \subset M \times R$ with $(R^1, D_0) \cong R_1$ and $N_1 + R_1 = M \times R$ and $N_1 \cap R_1 = \{0\}$. Let $x = (y,z) \in R_1$ be the image of $1 \in R$ under $(R^1, D_0) \cong R_1$. Since $D(1) = 0$, $D(x) = D((y,z)) = (0,0)$. As $D((y,z)) = (D_M(y), D_R(z))$, we have $D_R(z) = 0$, so $z \in C$, and $D_M(y) = 0$ also.

Suppose that $z = 0$. Since $R_1 = Rx = R(y,z)$, $R_1 = R(y,0) = (Ry,0) \subset M \times \{0\}$. This means that $(N_1 \cap (M \times \{0\}) + R_1 \subseteq M \times \{0\}$. Let $m \in M$. Like all elements of $M \times R$, we have $(m,0) = n_1 + r_1$ where $n_1 \in N_1$ and $r_1 \in R_1$ are unique. Moreover, $r_1 = rx = (ry,0)$ for some $r \in R$. So $(m,0) = n_1 + (ry,0)$ which implies $n_1 = (m - ry, 0)$. This means that $n_1 \in N_1 \cap (M \times \{0\})$. So $(m,0) = n_1 + r_1 \in (N_1 \cap (M \times \{0\}) + R_1$. It follows that $M \times \{0\} = (N_1 \cap (M \times \{0\}) + R_1$. Since $N_1 \cap R_1 = \{0\}$ this sum is direct. Let $N_2 = N_1 \cap (M \otimes \{0\})$. Thus $M \times \{0\} = N_2 \oplus R_1$. Abstractly, we have a differential isomorphism $M \cong N_2 \oplus (R^1, 0) = N_2 \oplus (\{0\} \times R)$. Now $M \times R = (M \times \{0\}) \oplus (\{0\} \times R)$ so $M \times R = (N_2 \oplus R_1) \oplus (\{0\} \times R)$. And we also have $M \times R = N_1 \oplus R_1$. All these direct sums are differential. Using the first, we have $(M \times R)/R_1 \cong (N_2 \oplus (\{0\} \times R)$ and the second $(M \times R)/R_1 \cong N_1$ so $N_2 \oplus (\{0\} \times R) \cong N_1$. Since $M \cong N_2 \oplus (\{0\} \times R)$ we conclude that $M \cong N_1$.

Now suppose that $z \neq 0$. Since C is a field, z has an inverse. Then $R_1 = Rx = Rz^{-1}x$. Let $w = z^{-1}y$, so $z^{-1}x = (w,1)$. Note that $D(w) = 0$. Consider $T : M \times R \to M \times R$ defined by $T((m,r)) = (m + rw, r)$. T is a differential endomorphism, indeed an isomorphism with inverse S where $S((m,r)) = (m - rw, r)$. Then $T((0,1)) = (w,1)$ so $T(R(0,1)) = R(w,1) = R_1$ and $S(R_1) = R(0,1)$. T is the identity on $(M,0)$. Let $N = S(N_1)$, so $M \times R = S(M \times R) = S(N_1 + R_1) = S(N_1) + S(R_1) = N + R(0,1)$. Thus for any $(m,r) \in M \times R$ there are $n \in N$ and $s \in R$ such that $(m,r) = n + s(0,1)$, so $n = (m, r - s)$. This means that every element of M occurs as the first entry of some element of N. As S is an isomorphism, $N_1 \cap R_1 = \{0\}$ implies $S(N_1) \cap S(R_1) = \{0\}$. Thus $N \cap R(1,0) = \{0\}$, which means that if $(0,r) \in N$ then $r = 0$. If two elements of N have the same first entry, say (m,t) and (m,u) then since $(0, t-u) = (m,t) - m(u) \in N$, $t - u = 0$ and $t = u$. Thus there is a function $f : M \to R$ such that $N = \{(m, f(m)|m \in M\}$. Since N is a differential R submodule of $M \times R$, $f \in \mathrm{Hom}_R^{\mathrm{diff}}(M,R)$. Then $M \to N$ by $m \mapsto)m, f(m))$ is a differential homomorphism, in fact a differential isomorphism with inverse $(m, f(m)) \mapsto m$. So $M \cong N$. Since $N \cong N_1$ usiing S, we conclude that $M \cong N_1$ as desired.

The assumption that R^D is a field also implies that $(R^1, 0)$ can appear as a direct summand in a direct sum if and only if it appears as a direct summand in one of the summands:

Lemma 5.5 *Suppose R^D is a field. Suppose (V, D_V) and (W, D_W) are differential finitely generated R modules such that $(W, D_W) \oplus (V, D_V)$ has a differential direct summand isomorphic to (R, D_0). Then either*

(W, D_W) or (V, D_V) has a differential direct summand isomorphic to (R^1, D_0).

Proof. Under the assumptions given, there are differential homomorphisms $\phi : (R^1, D_0) \to (W, D_W) \oplus (V, D_V)$ and $\psi : (W, D_W) \oplus (V, D_V) \to (R^1, D_0)$ such that $\psi\phi$ is the identity. Let $f_W : (W, D_W) \to (R^1, D_0)$ (respectively $f_V : (V, D_V) \to (R^1, D_0)$ be ψ preceded by inclusion on the first (respectively second) summand, and $g_W : (R^1, D_0) \to (W, D_W)$ (respectively $g_V : (R^1, D_0) \to (V, D_V)$) be ϕ followed by projection on the first (respectively second) summand. For $r \in (R^1, D_0)$, $r = \psi(\phi(r)) = f_W(g_W(r)) + f_V(g_V(r))$. In particular, either $f_W g_W$ or $f_V g_V$ is non-zero. Since $f_W g_W$ (respectively $f_V g_V$) is a differential endomorphism of (R^1, D_0) which is hence given by multiplication by a constant c_W (respectively c_V), at least one of c_W, c_V is non-zero. Say $c_V \neq 0$. Since R^D is a field, c_V is invertible. Let $h_V : (V, D_V) \to (R^1, D_0)$ be $h_V = c_V^{-1} f_V$. Then $h_V g_V$ is the identity, which shows that $g_V : (R^1, D_0) \to (V, D_V)$ is differential injection with left inverse h_V, so that $g_V((R^1, D_0))$ is a differential direct summand of (V, D_V) isomorphic to (R^1, D_0).

Whether R has free cancellation or not, for any module we can separate out the direct summands which are candidates for cancellation. Call a differential finitely generated projective R module module *trivial free* if it has no differential direct summands isomorphic to (R^1, D_0). Assuming, as we do, that a differential finitely generated projective module P has finite rank, we can start peeling off differential direct summand summands until we have $P = C \oplus (R^n, D_0)$ where C is trivial free. We call C a *core* of P. Note that $(P) = (C)$ in $PC_0^{\text{diff}}(R)$. Let $TF(R) = \{[M] \in M^{\text{diff}}(R) | M \text{ is trivial free}\}$. We have just remarked that every differential finitely generated projective R module V is a direct sum of a trivial free module and a module of the form (R^n, D_0). Thus $M^{\text{diff}}(R) = TF(R) + F_0^{\text{diff}}(R)$. Moreover $TF(R) \to PC_0^{\text{diff}}(R)$ is surjective.

Note that if R^D is a field the contrapositive of Lemma 5.5 says that the direct sum of trivial free modules is trivial free. This means that $TF(R) \subset M^{\text{diff}}(R)$ is a submonoid.

There is no reason for a core of P, or the complementary summand, to be unique up to isomorphism. If R has free cancellation, this ambiguity disappears.

Lemma 5.6 *Let R have free cancellation. Suppose (V, D_V) and (W, D_W) are trivial free differential finitely generated projective modules and m, n are non-negative integers such that $(V, D_V) \oplus (R^n, D_0) \cong (W, D_V) \oplus (R^m, D_0)$. Then $m = n$ and $(V, D_V) \cong (W, D_W)$.*

Proof. We may assume that $m \geq n$. Then free cancellation implies that $(V, D_V) \cong (W, D_W) \oplus (R^{m-n}, D_0)$. Since (V, D_V) is assumed to be trivial free, $m - n = 0$, and then both conclusions follow.

Lemma 5.6 implies that, under free cancellation the isomorphism class of the core of a differential finitely generated projective R module is determined by the module: if (P, D_P) is a differential finitely generated projective R module with cores (V, D_V) and (W, D_W) then $(P, D_P) = (V, D_V) \oplus (R^n, D_0)$ and $(P, D_P) = (W, D_V) \oplus (R^m, D_0)$ so (under free cancellation) $(V, D_V) \cong (W, D_W)$. For a differential finitely generated projective R module P, we write $\mathcal{C}(P)$ for a core of P (equivalently, a representative of the isomorphism class of all cores of P). If P is trivial free, then P is a core of itself; in this case, we choose $\mathcal{C}(P) = P$. At the other extreme, $\{0\}$ is a core of (R^n, D_0) and we choose $\mathcal{C}((R^n, D_0)) = \{0\}$.

When R^D is a field then R has free cancellation, so by Lemma 5.6, differential finitely generated projective modules have cores, and by (the contrapositive of) Lemma 5.5, the direct sum of trivial free modules is trivial free. This implies that the core is additive:

Corollary 5.6 *Assume R^D is a field. Let V and W be differential finitely generated projective R modules. Then $\mathcal{C}(V) \oplus \mathcal{C}(W) \cong \mathcal{C}(V \oplus W)$.*

Proof. We have $V \cong \mathcal{C}(V) \oplus (R^n, D_0)$ and $W \cong \mathcal{C}(W) \oplus (R^m, D_0)$ for some n, m. Then $V \oplus W \cong \mathcal{C}(V) \oplus \mathcal{C})W) \oplus (R^{n+m}, D_0)$. Since $\mathcal{C}(V) \oplus \mathcal{C}(W)$ is trivial free, this last isomorphism shows that $\mathcal{C}(V) \oplus \mathcal{C}(W)$ is a core of $V \oplus W$.

We already know that $TF(R) \to PC_0^{\text{diff}}(R)$ is surjective. If R^D is a field, in which case $TF(R)$ is a monoid, it is actually an isomorphism.

Theorem 5.1 *Suppose R^D is a field. Then the monoid homomorphism $TF(R) \to PC_0^{diff}(R)$ by $[V] \mapsto (V)$ is an isomorphism.*

Proof. We already know the map is a surjective monoid homomorphism. Suppose V and W are trivial free differential finitely generated projective R modules such that $(V) = (W)$ in $PC_0^{\text{diff}}(R)$. Then for some n, m, $V \oplus (R^n, D_0) \cong W \oplus (R^m, D_0)$. By Proposition 5.8 R has free cancellation. By Lemma 5.6 $V \cong W$. Thus $[V] = [W]$, showing injectivity.

When R^D is a field, for a differential finitely generated projective R module V, write $c(V)$ for $\text{rank}(V) - \text{rank}(\mathcal{C}(V))$, so $V \cong \mathcal{C}(V) \oplus (R^{c(v)}, D_0)$. Then $[V] \mapsto [\mathcal{C}(V)], ((R^{c(V)}, D_0))$ gives a monoid isomorphism $M^{\text{diff}}(R) \to TF(R) \times F_0^{\text{diff}}(R)$ and then by Theorem 5.1, a monoid isomorphism $M^{\text{diff}}(R) \cong PC_0^{\text{diff}}(R) \times F_0^{\text{diff}}(R)$. We can view this last isomorphism as saying that nothing about the nature of the classification of differential finitely generated projective R modules up to isomorphism is lost by passing to the coarser classification up to classes in $PC_0^{\text{diff}}(R)$ – at least in the case R^D is a field.

Index